Cases on Error Analysis in Foreign Language Technical Writing

Nadezhda Anatolievna Lebedeva
Kherson State Agrarian and Economic University, Ukraine

A volume in the Advances in Educational Technologies and Instructional Design (AETID) Book Series

IGI Global
PUBLISHER of TIMELY KNOWLEDGE

Published in the United States of America by
IGI Global
Information Science Reference (an imprint of IGI Global)
701 E. Chocolate Avenue
Hershey PA, USA 17033
Tel: 717-533-8845
Fax: 717-533-8661
E-mail: cust@igi-global.com
Web site: http://www.igi-global.com

Library of Congress Cataloging-in-Publication Data

Names: Lebedeva, Nadezhda, 1974- editor.
Title: Cases on error analysis in foreign language technical writing / edited by Nadezhda Lebedeva.
Description: Hershey, PA : Information Science Reference, [2023] | Includes bibliographical references and index. | Summary: "This book will aim to provide relevant theoretical foundations and the latest empirical research findings in the area of error analysis in FL Technical writing. This project is also intended to summarize the experience in teaching a foreign language for professional purposes in various fields of higher education. We will seek specific cases in the classification of the errors committed and, specifically, the treatment given to such errors by specialists from all over the world. Different classifications of error analysis will be also welcomed. This implies to gather cases from many different fields regarding technical writing in a foreign language for special purposes"-- Provided by publisher.
Identifiers: LCCN 2023003354 (print) | LCCN 2023003355 (ebook) | ISBN 9781668462225 (hardcover) | ISBN 9781668462263 (paperback) | ISBN 9781668462232 (ebook)
Subjects: LCSH: Language and languages--Study and teaching--Error analysis. | Academic writing--Study and teaching (Higher) | Technical writing--Study and teaching (Higher) | LCGFT: Essays.
Classification: LCC P53.3 .C37 2023 (print) | LCC P53.3 (ebook) | DDC 418.0071/1--dc23/eng/20230314
LC record available at https://lccn.loc.gov/2023003354
LC ebook record available at https://lccn.loc.gov/2023003355

This book is published in the IGI Global book series Advances in Educational Technologies and Instructional Design (AETID) (ISSN: 2326-8905; eISSN: 2326-8913)

British Cataloguing in Publication Data
A Cataloguing in Publication record for this book is available from the British Library.
All work contributed to this book is new, previously-unpublished material.
The views expressed in this book are those of the authors, but not necessarily of the publisher.
For electronic access to this publication, please contact: eresources@igi-global.com.

Advances in Educational Technologies and Instructional Design (AETID) Book Series

ISSN:2326-8905
EISSN:2326-8913

Editor-in-Chief: Lawrence A. Tomei, Robert Morris University, USA

MISSION

Education has undergone, and continues to undergo, immense changes in the way it is enacted and distributed to both child and adult learners. In modern education, the traditional classroom learning experience has evolved to include technological resources and to provide online classroom opportunities to students of all ages regardless of their geographical locations. From distance education, Massive-Open-Online-Courses (MOOCs), and electronic tablets in the classroom, technology is now an integral part of learning and is also affecting the way educators communicate information to students.

The **Advances in Educational Technologies & Instructional Design (AETID) Book Series** explores new research and theories for facilitating learning and improving educational performance utilizing technological processes and resources. The series examines technologies that can be integrated into K-12 classrooms to improve skills and learning abilities in all subjects including STEM education and language learning. Additionally, it studies the emergence of fully online classrooms for young and adult learners alike, and the communication and accountability challenges that can arise. Trending topics that are covered include adaptive learning, game-based learning, virtual school environments, and social media effects. School administrators, educators, academicians, researchers, and students will find this series to be an excellent resource for the effective design and implementation of learning technologies in their classes.

COVERAGE

- K-12 Educational Technologies
- Bring-Your-Own-Device
- Web 2.0 and Education
- Game-Based Learning
- Higher Education Technologies
- Social Media Effects on Education
- Adaptive Learning
- Digital Divide in Education
- E-Learning
- Curriculum development

IGI Global is currently accepting manuscripts for publication within this series. To submit a proposal for a volume in this series, please contact our Acquisition Editors at Acquisitions@igi-global.com or visit: http://www.igi-global.com/publish/.

The Advances in Educational Technologies and Instructional Design (AETID) Book Series (ISSN 2326-8905) is published by IGI Global, 701 E. Chocolate Avenue, Hershey, PA 17033-1240, USA, www.igi-global.com. This series is composed of titles available for purchase individually; each title is edited to be contextually exclusive from any other title within the series. For pricing and ordering information please visit http://www.igi-global.com/book-series/advances-educational-technologies-instructional-design/73678. Postmaster: Send all address changes to above address.

Titles in this Series

For a list of additional titles in this series, please visit:
http://www.igi-global.com/book-series/advances-educational-technologies-instructional-design/73678

Phenomenological Studies in Education
Jason D. DeHart (University of Tennessee, Knoxville, USA)
Information Science Reference • © 2023 • 328pp • H/C (ISBN: 9781668482766) • US $215.00

Implementing Rapid E-Learning Through Interactive Materials Development
Mohammad Issack Santally (University of Mauritius, Mauritius) Yousra Rajabalee (Mauritius Institute of Education, Mauritius) and Ravi Rajputh (University of Mauritius, Mauritius)
Information Science Reference • © 2023 • 300pp • H/C (ISBN: 9781668449400) • US $215.00

Practices That Promote Innovation for Talented Students
Julia Nyberg (Purdue University Global, USA) and Jessica Manzone (University of Southern California, USA)
Information Science Reference • © 2023 • 300pp • H/C (ISBN: 9781668458068) • US $215.00

Using Assistive Technology for Inclusive Learning in K-12 Classrooms
Jeremy Bell (University of Louisiana at Monroe, USA) and Timothy Gifford (Movia Robotics Inc, USA)
Information Science Reference • © 2023 • 300pp • H/C (ISBN: 9781668464243) • US $215.00

STEM Education Approaches and Challenges in the MENA Region
Fatimah Alhashem (Gulf University for Science and Technology, Kuwait) Heather Pacheco-Guffrey (Bridgewater State University, USA) and Jacquelynne Boivin (Bridgewater State University, USA)
Information Science Reference • © 2023 • 300pp • H/C (ISBN: 9781668468838) • US $215.00

For an entire list of titles in this series, please visit:
http://www.igi-global.com/book-series/advances-educational-technologies-instructional-design/73678

IGI Global
PUBLISHER of TIMELY KNOWLEDGE

701 East Chocolate Avenue, Hershey, PA 17033, USA
Tel: 717-533-8845 x100 • Fax: 717-533-8661
E-Mail: cust@igi-global.com • www.igi-global.com

Editorial Advisory Board

Table of Contents

Detailed Table of Contents

coalescence of these three domains, which is natural yet rare in research. Data collection tools included interviews, document analysis, and random participant observation. Error analysis was also used as a methodological approach to examine longitudinal data from a cohort of B1-level English language learners. The initial technical writing work revealed a failure to adopt the nature, structure, and content of technical writing. Yet, later-produced works exhibited better development over time after the interventions, which were full of specific strategies. By validating the synergy in error correction between process writing, which clears clouded minds with its precise step-by-step layout, and social constructivism, which helps generate meaning and cooperate, the study concludes the necessity of such integration into technical writing classes.

Chapter 3
Error Analysis of Teacher Technical Writing: The Corpus of EFL Lesson Plans54
Banu Çiçek Başaran Uysal, Eskisehir Osmangazi University, Turkey

This chapter delves into an investigation of pre-service English as a foreign language teachers' technical writing products, by employing the Surface Strategy Taxonomy proposed by Dulay et al. (1982). By shedding light on the written errors of future English language teachers, this study contributes to our understanding of their language learning needs and presents implications for improving teacher education practices. The chapter examines a corpus of lesson plans written by senior students (n=140) enrolled at a public university, aiming to become English language teachers. Lesson planning can be considered a form of technical writing for language teachers since it requires adherence to specific techniques and the use of appropriate language structures to effectively convey in-class procedures. The analysis reveals that the prospective teachers produced erroneous utterances falling into four distinct categories: omission (45%), misinformation (30%), addition (17%), and misordering (7%). These findings serve as a foundation for several suggestions to enhance teacher education.

Chapter 4
Errors in the Written Productions of Spanish Undergraduate Engineering Students: A Comparative Analysis 74
Antonio Daniel Juan Rubio, Universidad de Granada, Spain

Writing is a language skill that for most FL students is difficult to master. Therefore, students often find it difficult to write a specific composition which is not their FL. Especially for undergraduate technical students, this seems to be a challenging task since the presence of difficulties in any FL writing forces them to commit errors in their written productions. Therefore, what this chapter will identify is those errors

made by engineering students in a public Spanish Polytechnic University over the last two years. The authors shall attempt to categorise errors based on their source following the comparative taxonomy proposed by Dulay, Burt, and Krashen. So, a total of 60 essays undertaken by these engineering students were selected taking into account their productions in the task assigned during the semester. The main objective is to prove that the results have undoubtedly demonstrated that the majority of errors committed by these engineering students can be classified as belonging to the interlingual category, being the category with the highest number of occurrences.

Architecture is one of the fundamental and promising specialties of a modern agricultural university. It is impossible to imagine the study of world experience in this area without ESP since the advanced design literature is written in English. Writing, as a skill that involves all the senses, is a fundamental means of teaching future architects in a foreign language class. A lot of time is devoted to the study of a professional foreign language in the educational programs of agricultural universities. The purpose of this chapter is a theoretical description of the practical experience of teaching ESP writing to future architects of an agricultural university. The novelty of the work is to describe the Ukrainian experience of teaching writing in a specialized foreign language at the Faculty of Civil Engineering of the Agricultural University based on scientific sources and existing programs.

The features of teaching a foreign language to lecturers of the higher school of non-linguistic specialties are considered. Relevant competencies are proposed. The features of the scientific text as the main product of the work are considered. A system of tasks is proposed in the context of simultaneous teaching of a foreign language and deepening of professional scientific competencies. Examples are given, including the development of an educational and methodological complex of the discipline "Land Management" (Science and Land). The unique experience of using modern foreign language teaching technologies in North Korea is described.

Preface

There is a clear tendency to consider the errors committed by learners in the process of learning a foreign language not as a negative aspect but as a natural process in the development of their language skills. In the past, teachers and researchers pondered these errors as something unfavourable, something to prevent from occurring at all costs. To regulate this appearance of errors, many researchers in the field of foreign language technical teaching have started to study and track the errors made by these students. Knowing the most common types of errors is an important and necessary step to inform teachers about where to direct their efforts regarding the most problematic areas students face while producing their technical writing assignments in a foreign language.

Cases on Error Analysis in Foreign Language Technical Writing provides relevant theoretical foundations and the latest empirical research findings in the area of error analysis in foreign language technical writing. This book also summarizes the experience of teaching a foreign language for professional purposes in various fields of higher education. Covering topics of the book are such as error analysis, lexical skills, writing outputs, foreign language for specific purposes and so on. This casebook is an essential resource for preservice teachers, educators, administrators, librarians, researchers, and academicians.

The main aim of this book is to provide the relevant theoretical background and the latest results of empirical research in the field of error analysis in technical writing. Technical writing is interpreted as descriptive writing, which separates the author's opinion from statistical facts, causal relationships and thematic evidence. This project is also intended to summarize the experience of teaching a foreign language for professional purposes in various areas of higher education. Specific cases were found in the classification of the mistakes made. The interpretation of such errors will become useful and interesting for specialists all over the world. The target audience of the book consists of professionals and researchers working in the field of error analysis in technical writing in a foreign language, teachers of non-linguistic higher educational institutions, and technical and other specialities. In addition, the book will provide readers with information from many of the special cases associated with teaching technical writing in a foreign language. The chapters presented in the book are related to, but not limited to, the following topics: teacher training, higher education, language studies, foreign language teaching, foreign language technical writing, foreign language for environmental, agricultural and physical sciences, foreign language for engineering. Teachers will use the results of the error analysis to help students improve their writing skills and reduce errors. Needless to say, writing is a language skill that for most foreign language students is quite difficult to master. Especially for undergraduate technical students, this seems to be a challenging task since the presence of difficulties in any foreign language technical writing forces them to commit various errors in their written productions. This book can be roughly divided into two parts: the first presents a study of the mistakes made by students of higher educational institutions. The second part is devoted more to the problems of learning a written foreign language by students of non-linguistic specialities of a narrow profile - a foreign language for specialized purposes.

So, in the first part of the book, researchers from around the world are offered the opportunity to recognize and understand the main causes and types of forms of the language being studied in the process of classifying errors and rules. As it is known, error analysis is a branch of applied linguistics that studies the mistakes made by students, and also reflects certain universal strategies for overcoming them. The trend in the study of errors suggested certain reactions to contrast analysis, in which the phenomenon of interference is considered the main source in the study of any second language. Thus, the errors have attracted the great attention of scientists and researchers who have tried to analyze their diversity. As a consequence, different error analysis procedures have been introduced from very different and specific points of view of scientists. Based on the published research results of the authors of this book, taxonomies and strategies for effective foreign language teaching are proposed with recommendations for the construction of appropriate curricula. Encouraging users to use language more accurately to practice targeted communication conditions the consideration of errors as a guide to learning and indicators of its control. The use of taxonomy helps researchers identify problem areas in interlingual communication. Error analysis and categorization focus on material data, namely the actual use of the language. Determination of the main causes of errors required the study of internal and external factors in the process of learning a foreign language. Accordingly, the book aims to provide an analysis of the technical writing errors of a wide range of professionals and to provide insight into the most common errors. This huge collective work is a serious attempt to give an insight into the context of teaching and learning a foreign language in universities where participants receive higher professional training. The expanding range of opportunities in the professional field provides a unique context for English learners compared to the internal and external limitations of opportunities to obtain professional data in the target language. The book contains the necessary research to understand the unique challenges and opportunities that arise in the context of an expanding circle and how this affects language learning and the learning of foreign language learners. The critical role of writing effective English language teaching plans is emphasized, as the need for teachers to develop their language skills to be effective in their profession.

The results published in the book highlight the importance of writing in improving the quality of language teaching and learning. By identifying the most common types of errors, understanding these errors can broaden the field for developing more effective curricula, and helping students improve their language skills more effectively. The results published in the book can be used to develop and improve training programs for English teachers for undergraduate and graduate students in construction, agriculture and other areas. Based on published results, it has been observed in the field of English learning that it would be beneficial to receive additional training in target language grammar for non-linguistic students. Specialist training programs may include grammar courses, including technical writing exercises and self-correction techniques in the curriculum. The book shows quite extensively the teaching methods that are best suited for students. For teachers of a foreign language, the book is also valuable in that they can evaluate their learning strategies and identify areas for their development, focusing on knowledge of students' mistakes. In this way, teaching methods become more fun, and new research knowledge is useful for the professional development of teachers. Based on the results of the book, it is shown that professional development programs can include professional English language courses for teachers, help them identify and correct their mistakes through outreach activities, and can offer a network to share feedback on their written work. Error analysis can also help educators better understand the difficulties their students face. Teachers will be able to improve the learning environment by understanding the nature of student errors. In addition, curriculum developers can add book content to their courses to improve English teachers' technical writing practices before they begin, or components of existing courses can be revised to include tasks related to the most commonly observed errors. Despite its limitations, the present study certainly improves the scientific understanding of technical writing errors by building on the main categories of student error.

Many error taxonomies are based on the linguistic element that the error affects. These linguistic taxonomies classify errors according to the language component and the linguistic component affected by the error. Errors can be classified into different taxonomies. Poor English proficiency in terms of writing skills may well be the cause of ambiguous errors. Errors come from sources such as intralingual and cross-lingual and must be corrected properly. In general, error analysis associated with process-oriented writing and a socially constructive approach contributes greatly to student learning. The process of assimilation of special words and structures is faster, and writing skills are strengthened. Writing is a very important component in the learning process because the memory of the written expression of one's thoughts is important not only in itself but also as a tribute to the promotion of spiritual thoughts. The use of auditory-motor images of internal movement and visual-motor movements in the process of writing provides a comprehensive assimilation of knowledge. There is also an emphasis on conducting a comparative evaluation analysis (independently or under the guidance of a teacher) of native and foreign texts. The texts are accompanied by lexico-stylistic and translation-methodological commentary, which provides the appropriate justifications, admissible options and appropriate references to the theoretical part (where necessary). Writing works with technical texts is a conscious action from the very beginning. All elements, forms and means used by students studying technical specialities, through written works, convey the desired content, which is the subject of awareness. The use of lexical units of profile topics further increases the level of awareness.

Graphic images of words and structural patterns of sentences learned through written exercises are more easily recognized by visual perception. In a written speech, the speech character of a coherent statement, usually consisting of several logically and structurally interconnected sentences, is revealed more clearly. So, in the process of teaching foreign language communication to students, an important function of the analysis of writing errors is that it can serve as an effective means of control. Written work, both current and final, is used to test knowledge of the language material, and practical skills in listening, speaking and reading. Current tests using technical writing are conducted after mastering a certain part of the language material or oral topic. Writing is an important means of improving the lexical and grammatical skills of students specializing in various areas of modern knowledge. The development of modern science and education stimulates the acceleration of the exchange of information between their participants - researchers, teachers of higher educational institutions and students. Writing provides an opportunity to exchange and acquire knowledge, skills and abilities. The second part of the book contains information related not only to a specialized written foreign language, but the value lies in the fact that it touches upon the problems of intercultural communication, and also provides extensive practical material for teaching teachers of narrow technical specialities who do not speak English. Intercultural communication between teachers and foreign-speaking students is carried out using written English. The problem of implementing joint programs of various foreign universities was solved thanks to the developments that were created during the Covid-19 pandemic. Knowledge of professional English is necessary for teachers of foreign universities and is provided during training in the magistracy and postgraduate studies. In the modern world, a foreign language plays the role of a necessary means of communication between teachers and scientists from different countries. A specialist who speaks little English either deliberately puts himself on the periphery of the scientific field or must make efforts to achieve an acceptable level (fluent reading without a dictionary). As a rule, mastering and/or improving the level of foreign language proficiency occurs in parallel with the development of scientific research. A specialist, as a rule, is critical and attentive to the acquired knowledge, evaluating them primarily from an applied point of view. He understands that he urgently needs, first of all, a certain minimum to enter the circle of colleagues who communicate in another language; that mastering the language is not the goal of his efforts, and that the ability to grasp the essence of a foreign text on the fly will save much time and effort in the future in selecting the necessary information. A foreign language for special purposes is developed to meet the specific needs of students in narrow specialities and is intended for specific disciplines. It is directly related to the methodology and activities of the discipline it serves and focuses on the special vocabulary and genre of this discipline. A foreign language for special purposes is intended for adult learners in a professional work activity. The need for theoretical development of the issues of teaching non-linguistic specialities a professional foreign language, the search for effective means of teaching oral and written foreign language communication, the poor elaboration of the issues of teaching university teachers a professional foreign language - all these problems are raised in the second part of this scientific book. A ready-made program for training is given, which can be used in the pedagogical process in any country in the world. The book has a narrow specialized focus, which is its special value. However, the circle of readers of this work seems to us the widest precisely because of the practical developments of the book, its ready-made programs and its texts. A foreign language for specialized purposes is directly related to the methodology and activities of the served discipline, focusing on special vocabulary and the genre of technical specialization. A foreign language for special purposes is intended for adults studying professional work. Needless to say, writing is a language skill that is quite difficult for most foreign language learners to master. Especially for students, this seems to be a difficult task, since the presence of difficulties in technical writing in any foreign language forces them to make various mistakes in their written work. This is important for developing different teaching strategies that can minimize student errors.

After analyzing scientific sources and summarizing their own experience in teaching a written foreign language of a profile orientation, the authors of this book came to valuable conclusions and provided further practical recommendations and tasks for widespread use by foreign colleagues. As you know, undergraduate and graduate students master the culture of professional speech with the help of written speech. Possession of written competence in a foreign language contributes to the development of professional qualities of specialists in technical specialities, to an increase in the level of general culture, their outlook, as well as intelligence. Proficiency in a written foreign language allows you to move up the career ladder in the future. By improving the written language, new words are memorized and vocabulary is enriched. Students of technical specialities need to understand that mastering writing is very important, along with other language and communication competencies. Writing skills open access to new information and educational resources and intersect with creative thinking. However, speech and grammatical errors are necessary accompanying steps in learning any language. To avoid errors, a thorough study of the creation of written scientific texts in two languages is suggested. Here the logical structure, style and grammar are taken into account, special attention is paid to special vocabulary. Writing is the skill of creating full-fledged scientific papers that will later present the results of their research and meet the requirements for scientific papers around the world. Teaching writing remains a relevant and effective method, despite the abundance of innovative teaching aids, as it involves the visual, tactile and auditory, mental receptors of a person and contributes to the high-quality memorization of a specialized language, as noted above. Another aspect of the problem under study is the additional education of teachers. People from different countries and nationalities have a need and opportunity for direct intercultural communication. The current level of development of technologies and technical means can satisfy this social demand. The ability to write correctly in a foreign language allows a specialist to get involved in the work of attracting additional foreign scientific resources to their research.

The training of teachers of higher educational institutions of non-linguistic specialities in a foreign language has not yet been practically developed, since the main interest of researchers is focused on students. Since this scientific work presents research by scientists from Spain, Turkey, Ukraine and other countries, the value of this book lies in the typological comparison of languages, which is realized through teaching English as a foreign language. Contrastive analysis considers the influence of the native language on the one being studied. Such language interaction is a source of errors in teaching writing specialized English. This approach to the problem does not lose its relevance and constantly attracts the attention of teachers and researchers. Features of the process of teaching English to teachers of non-linguistic specialities in terms of advanced training is a special and non-standard view of the methodology of teaching a foreign language of narrowly professional orientation. It describes the improvement of the professional training of teachers of non-linguistic education in the context of lifelong learning for successful integration into the international scientific and educational community. The experience of teaching foreign language communication to university teachers in technical specialities is summarized. The reliability of the results and conclusions of the book's research is substantiated by the theoretical and methodological study of the problems under consideration and the use of a combination of correct qualitative methods for analyzing the material. Thus, this book contains a colossal practical experience that will be of interest to a fairly wide range of readers.

In the first chapter, anyone interested in the interaction of Spanish and English will find comprehensive information about the analysis of the mistakes of Spanish-speaking students in learning English. The unique study "Analysis of Errors in the Written Essays of Engineering Students: Using a Comparative Taxonomy" was conducted by Dr. Isabel María García and represents the most interesting scientific work in the field of comparative taxonomy. The researcher collected 72 essays and carefully analyzed the mistakes made by the students. It is important to note that the work on the collection of materials and analysis lasted two years and was reflected in the corresponding graphs and tables, confirming the correctness of the conclusions about the ratio of Spanish and English in the field of narrow professional engineering use. Error analysis serves at least three purposes, implemented in the process of writing, and teaching a foreign language. It shows the degree of mastering the language material, the assimilation of information contained in statements in another language and presenting difficulties and the most effective way to master the knowledge of the language being studied. That is, the presence and analysis of errors are positive in the pedagogical process. The classification of errors is also an important aspect of the process of acquiring deep knowledge applied in the future profession. The author of the chapter reveals the features of the 4 main types of errors, describing each of them in detail. The study is a comprehensive analysis of the scientific literature and its practical application in the interpretation of the latest data obtained. Valuable recommendations are offered to Spanish-speaking students of technical specialities, during which their knowledge of a written foreign language will be improved.

In the second chapter, "From Error Analysis to Effective Writing Outputs: An EAP Process-Oriented and Social-Constructivist Perspective," Dr. Ali Kurt reveals promising approaches to the methodology of teaching a foreign language. The author of the chapter describes the features of teaching English for academic purposes and English specific purposes, focusing on the differences between these disciplines in the pedagogical process of higher education. Technical writing is becoming more and more important and in demand, showing positive results in the development of disciplines and programs for English academic purposes and English specific purposes. The author of the chapter focuses on modern process-oriented writing, which consists of five stages (planning, in which students determine what and how to write; draft, where the first draft is written; editing, during which people think and edit the article; and final revision, where the final copy is written for readers). In other words, in technical writing, interlingual interference is the interference of the native language, which is the main obstacle, since it can complicate the process of ideological organization of text arrangement along with developed grammatical and lexical competence. In this case, the errors are systematic and force students to jump to conclusions. It is important to remember that, since the student has not yet mastered a foreign language, a special set of rules, principles and regulations, cross-language work, leading to errors, is a natural process of learning a foreign language, just like it happens when children learn a foreign language together with their native language.

The purpose of the second chapter of the book is to explore whether error analysis, developed for process-oriented writing classes within a social constructivist approach, can help second language learners improve their technical writing skills. Technical writing is a difficult task in the process of mastering a foreign language and is not considered acceptable for everyone, as it requires close attention to factual information, clarity, accuracy, well-thought-out structure, specific genre requirements and efficiency. Writing requires the use of an extended vocabulary and knowledge of the difficulties of applying grammar rules. Understanding and explaining the complex information and terms needed to complete tasks is also difficult for non-specialists. As a result, technical tasks require special attention, both from teachers and students. For reasons intended to facilitate technical writing, this request is relevant, interrelated, and significant in terms of research time and effort. In the opinion of students, the performance of technical written tasks, such as writing instructions, is of an appropriate and satisfactory standard. The main goal of the research described in this chapter is to develop ways to facilitate the process of planning and writing technical writing and alleviate complaints about the complexity of technical writing tasks while contributing to the methodological literature. This study identifies and explains errors in technical writing and combines the best features of relational and explanatory case studies. The concept of interest in the study is the language development of learners in second language acquisition to fill a gap in the existing literature, as no previous research has explored the confluence of these three areas.

A valuable experiment conducted by the author is described, which provides a unique experience for colleagues from other countries. A certain number of experimental samples also saturated the study with additional information, since the sample explained, and confirmed evidence and arguments quite a few times.

With the expansion of the present study, the scientist developed a 5-week program dedicated to technical writing, its main characteristics and the most common types. Materials corresponding to each type were initially presented to the class with writing samples. Using the text layout provided, a representation of related grammatical topics and vocabulary was created in a deductive way. The study was aimed at analyzing errors related to grammar, vocabulary, content, discourse and style. The main types of errors discussed for each category represent very valuable practical work. As a result, every grammatical error has been proven to weaken the overall impact of the written work, so second language learners and their teachers need to pinpoint the causes of errors in the first place. This is called error analysis, a process-oriented method that helps foreign language learners recognize, evaluate and classify the deviations that occur between their use of the language and the norms of the target language. The analysis can be done by the student himself or by a trained professional.

The study "Error Analysis of Teacher Technical Writing: The Corpus of EFL Lesson Plans" by Dr. Banu Çiçek Başaran Uysal is a comprehensive scientific work in the field of error analysis of teaching technical writing, which includes developed lesson plans. Language learner errors are defined as deviations from the norms of the target language. The author of the chapter talks about the context of language learning and the need to separate mistakes because they are different. Some are related to the lack of knowledge of students, and the second relates to deviations in academic performance due to inattention and fatigue. The author of the chapter considered the proposed practical determinations of whether a malformed statement is erroneous, taking into account the frequency of errors in the work of students. If correct forms are sometimes used and deviant forms are sometimes used, then it is possible to classify the error as such. On the other hand, if the wrong form is used all the time, then it is a mistake. Language learner errors are systematic and thus indicate progress in learning the target language. It is extremely important to study the technical errors in the writing of prospective teachers of English as a foreign language, as these errors indicate the level of language proficiency. The analysis of lesson plans is important for understanding the language needs of prospective English teachers because it is the understanding of language competence that is needed to conclude knowledge of the subject. Analysis of plans – these are the future teachers of English because they will teach children and students. Finding gaps in their knowledge is vital to teacher development and language planning. Therefore, the analysis of the mistakes of future specialists in the field of language teaching is crucial for further successful work.

According to the author of the chapter, creating lesson plans is technical writing for teachers, as they need to use a certain terminology and a certain set of grammatical structures to indicate classroom procedures and learning outcomes. The author notes how important the algorithm of the process of writing plans is, since in stages teachers should focus on the results, prior knowledge of language learners, expected problems and their solutions, models of interaction in the classroom, lesson stages, class management, the very procedure for teaching the team and the time of the activity. Lesson planning is not only about writing a series of steps to be completed in the same class, but also includes analyzing background and proficiency levels, finding and choosing suitable language teaching methods and activities to teach the topic, and synthesizing teaching theory and practice. The author of the chapter notes the importance of creating lesson plans as a roadmap for the learning process. After the lesson is over, the teacher can revise his plans and check the effectiveness of his actions, in accordance with which he can adjust his project. Because of the importance of lesson planning, future teachers are trained to write detailed and well-structured plans. Prospective teachers are trained during their undergraduate studies to write detailed and well-structured lesson plans, which demonstrates the improvement of technical writing skills and focus on the necessary elements of lesson design to ensure continued success.

In the fourth chapter, Dr. Antonio Daniel Juan Rubio in "Errors in the Written Productions of Spanish Undergraduate Engineering Students: A Comparative Analysis" gives extensive insight into the complexity of learning a foreign language. The author of the chapter considers error analysis as a natural step in the development of language skills, as the main element of the feedback system in the pedagogical process between its participants. Errors are seen as evidence of the creative process, and their analysis provides scientific evidence for the features of the language system used during the development of the learning process. The chapter focuses on four main steps: identifying errors, explaining errors, classifying them, and evaluating them. Errors that occur in the process of teaching students serve as a navigator for making the necessary adjustments to the learning process. The value of this chapter lies in its narrow focus: it is not working with university English students whose level of English is above average, but students of technical English in engineering whose initial level is much lower. All background information finds its natural reflection in the numerous mistakes made by students in written work. The research methodology consists of a comparative analysis of errors in the written work of engineering students over the last two academic years. The study gives a clear idea of the categories of errors and their frequency.

The fifth chapter, "ESP in the Field of Architecture: Features of Teaching Writing and Some Aspects of Error Analysis," by Nadezhda Anatolievna Lebedeva reveals the narrow specifics of a foreign language for architects and builders. A professional foreign language must be developed by mastering specialized vocabulary, the knowledge of which is unthinkable without such a serious learning tool as writing. It is written work that contributes to memorization and high-quality mastery of professional vocabulary. Future specialists in the field of architecture also acquire the necessary professional skills through writing. When mastering a specialized foreign language through writing, especially typical and practically significant texts on architecture and construction are taken from open sources on the Internet. Based on the fact that international contacts are expanding, the foreign language of architects is given increasing importance in agricultural universities. One of the most effective means of teaching writing in a specialized foreign language is translation despite the variety of methods, for the preparation of competitive specialists with broad career prospects and advancement opportunities. The chapter discusses popular methods of teaching specialized writing, describes their importance and effectiveness. Examples are given that can be used in the work of colleagues teaching at universities with similar specialties. Teaching writing in a foreign language of a specialized orientation to students of agricultural universities in construction specialities must necessarily include a written translation. Dictation is a useful and flexible type of activity in teaching a foreign language with an architectural orientation. These methods can be used when introducing a new structure, repeating and consolidating specialized vocabulary on architecture, and consolidating grammar. Dictation is an important and effective means of teaching writing in a foreign language, as well as a knowledge control tool. Mastering writing in a specialized foreign language are also annotations and summarizing. Teaching writing remains a relevant and effective method despite the abundance of innovative pedagogical aids since it involves the visual, tactile and auditory, mental receptors of a person and contributes to the high-quality memorization of a specialized language.

Olga Anatolievna Pasko dedicated her study “Features of Teaching a Foreign Language to Lecturers of Higher Educational Institutions” to such problem as teaching experienced tutors of non linguistic specialities. People from different countries and nationalities have a need and opportunity for direct intercultural communication. The current level of development of technologies and technical means can satisfy this social demand.

Proficiency in a written foreign language allows a specialist to get involved in the work of attracting additional foreign scientific resources to their research. This causes the need for additional training (independently or as part of advanced training). Currently, the methodological issues of teaching tutors a professional foreign language are practically not developed. But this chapter offers a comprehensive practical program for the improvement of teachers of narrow technical specialties. In the modern world, a foreign language plays the role of a necessary means of communication between teachers and scientists from different countries. A specialist who has little command of it either deliberately puts himself on the periphery of the scientific area, or must make efforts to achieve an acceptable level (fluent reading without a dictionary). As a rule, mastering the level of knowledge of a foreign language takes place in parallel with the development of scientific research, outside of working hours.

Therefore, a specialist, as a rule, is critical and attentive to the acquired knowledge, evaluating them primarily from an applied point of view. He understands that he urgently needs, first of all, a certain minimum to enter the circle of colleagues who communicate in another language; that mastery of the language is not the goal of his efforts, and that the ability to grasp the essence of a foreign text on the fly will save a lot of time and effort in the future for selecting the necessary information.

As a rule, during advanced training courses, students receive a lot of useful information about foreign-language resources of specialized literature, grants, research teams, etc. They write review abstracts in their speciality and learn to navigate the international scientific world. This directly helps in the preparation of one's publications and at the same time provides a strategy for the development of research. The chapter contains practical tasks, a thematic program that can be used in the educational process all over the world for specialists in "Land Management, Cadastre and Monitoring of Land".

The seventh chapter is devoted to "Improving the Lexical and Grammatical Skills of Students Specializing in Hydraulic Engineering and Water Resources by Means of Writing." It is an actual problem of modern education - the organizational and methodological foundations of writing and improving the lexical and grammatical skills of students of engineering specialties on the example of "Hydraulic engineering and water resources". To solve it, the author consistently solves several problems such as: considering the main methods and techniques of teaching a foreign language used in universities; formulating the linguo-didactic foundations for teaching English written speech in hydraulic engineering; developing criteria for selecting educational material for the formation of the competence of English-language written communication in hydraulic engineering (water resources).

As an object of research, the author chooses the process, and the subject – a methodology for teaching professionally oriented English-language written speech in the technical discipline. Skillfully and correctly the appropriate scientific methods – analytical, historical, and comparative are used in the chapter. The author shows that the effectiveness of teaching professionally oriented foreign language written communication in hydraulic engineering and water resources can be perfected by improving the lexical and grammatical skills of students, namely, expanding their horizons. communicative-linguistic component of professional competence in written communication in a foreign language through the phased use of exercises and tasks constructed system.

This research is distinguished by scientific novelty, deep theoretical study of the material and a wide scope of methodological application. The author competently combines the classical basic and modern innovative components of education. The material is structured clearly. The chapter's advantage is a logical and consistent presentation of the study and innovation in solving the problem of teaching professionally oriented English-language written communication at a high professional level.

The scientific study by Zinaida Lebiedieva "ESP Writing as a Practical Mean of Cultural Communication Between Ukrainian Music Teachers and Chinese Students" is devoted to a very important topic. It contains an interesting analytical review of modern sources of literature devoted to communication between Ukrainian music teachers and Chinese students, taking into account objectively established socio-cultural characteristics. The article has a high theoretical significance and practical value. The work is a complete study and does not need to be supplemented with information. The facts offered by the author are either taken from academic sources or correctly processed, so their reliability is beyond doubt. Moreover, the above research algorithm can be taken as a basis in similar studies between students and teachers of other language areas. The chapter pays worthy attention to the current situation of using extrasensory communication through writing and the prospects for its improvement. As the recent experience of the development of science and art under the conditions of covid restrictions has shown, this communication can become the leading one in the learning process. Therefore, its development is socially in demand and promising. This chapter is harmoniously balanced both in terms of the amount of information involved and in terms of the research methodology, the results obtained and the correctness of their interpretation. It will undoubtedly be of interest to the readership of this book, since it demonstrates the general and particular in teaching a foreign language - firstly, in different branches of knowledge, and secondly, by carriers of different cultures; thirdly, various forms of speech function (oral and written). References to literary sources are relevant, modern, valuable and significant in a scientific sense. The author develops existing knowledge and raises it to a new level. The study contains the author's recommendation, which allows other teachers to solve similar problems. In the context of increasing attention to the use of written language in the learning process, this enhances its relevance and prospects.

Nadezhda Anatolievna Lebedeva
Kherson State Agrarian and Economic University, Ukraine

Acknowledgment

Sincere gratitude to the IGI Global Publishing House and to all the authors who worked in writing the chapters of this scientific work.

Chapter 1
Analysis of Errors in the Written Essays of Engineering Students:
Using a Comparative Taxonomy

Isabel María García
Centro Universitario de la Defensa de San Javier, Spain

EXECUTIVE SUMMARY

There is a clear tendency to consider the errors made by learners in their process of language learning not as a negative aspect but as a natural step in the development of their language skills. In recent decades, researchers have considered errors as the evidence for a creative process in language learning. Therefore, the main objective of this chapter will be to classify the errors made by undergraduate engineering students in a public Spanish Polytechnic University over the last two courses. The authors categorise those errors following the comparative taxonomy proposed by Dulay, Burt, and Krashen in "Language Two" (1982). This taxonomy arranges errors into the following categories: interlingual, developmental, ambiguous, and other errors. Consequently, this chapter demonstrates that the most frequent category of errors is the interlingual category. And in order to lead this research, a corpus of a total of 72 essays was examined, comprising their written productions in the task assigned throughout these last two years.

DOI: 10.4018/978-1-6684-6222-5.ch001

INTRODUCTION

Nowadays there is a clear tendency to consider the errors made by learners in their process of language learning not as a negative aspect but as a natural step in the development of their language skills. In recent decades, researchers have considered errors as the evidence for a creative process in language learning. Therefore, the main objective of this paper will be to classify the errors made by undergraduate engineering students in a public Spanish Polytechnic University over the last two courses. We will try to categorise those errors following the comparative taxonomy proposed by Dulay, Burt, and Krashen in "Language Two" (1982). This taxonomy arranges errors into the following categories: interlingual, developmental, ambiguous, and other errors. Consequently, what we will demonstrate along this article is that the most frequent category of errors is the interlingual category. And in order to lead this research, a corpus of a total of 72 essays was examined, comprising their written productions in the task assigned throughout these last two years.

Error Analysis is a branch of Applied Linguistics which emerged in the 1960s to reveal that the errors that the learners did were not only due to their mother tongue but rather it also reflected certain universal strategies. This trend supposed a reaction to Contrastive Analysis that considered native language interference as the main source of errors in the field of second language learning.

Nzama (2010) states that error analysis is useful in second language learning because it reveals to the teachers, syllabus designers and textbook writers what the problem areas are. The teachers could design remedial exercises and focus more attention on the trouble shooting areas. Corder (1967, as cited in Nzama, 2010) states that errors are visible proof that learning is taking place. He has emphasized that errors, if studied systematically, can provide significant insights into how a language is actually learned by a foreigner. He also agrees that studying students errors of usage has immediate practical application for language teachers. In his view, errors provide feedback; they tell the teachers something about the effectiveness of his teaching. Sercombe (2000, as cited in Nzama, 2010) also explains that error analysis serves three purposes. Firstly, to find out the level of language proficiency the learner has reached. Secondly, to obtain information about common difficulties in language learning, and thirdly, to find out how people learn a language. From this statement it can be concluded that the study of errors should also be looked at as something positive both for learners and teachers

Jack C. Richards (1971) explained in that sense that "the field of errors analysis may be defined as dealing with the differences between the way people learning a language speak and the way that adult native speakers of the language use the language" (Richards, 1971: 1). Similarly, Corder (1981) stated "the errors that learners make are a major element in the feedback system of the process that we call language teaching and learning" (Corder, 1981: 35).

Since then, errors have received much attention from scholars and researchers who tried to analyse different errors made by language learners. Consequently, procedures have been introduced analysing errors from very different and specific points of view. Many error taxonomies have been based on the linguistic item that is being affected by the error. These linguistic taxonomies classify errors according to the language component and the linguistic constituent that the error affects.

Errors can be classified according to different taxonomies. The current research presented here follows the classification proposed by Dulay, Burt, and Krashen in "Language Two" (1982), being the surface strategy taxonomy and the comparative taxonomy two of the major ones in order to classify errors. We can also include two other classifications: linguistic category and communicative effect taxonomy. They firmly believed that an analysis of errors made by language learners could help us understand the process of language learning deeply.

Consequently, we will be basing ourselves on the analysis of errors according to the comparative taxonomy. The classification of errors following a comparative taxonomy is grounded on the comparison between the structure of L2 errors and other types of constructions. Therefore, this taxonomy includes four main categories of errors: developmental, interlingual, ambiguous, and other errors (arranged alphabetically).

Developmental errors are similar to those made by children learning the target language as their first language as in "cat eat it". *Interlingual errors* are similar in structure to a semantically equivalent phrase or sentence in the learners' native language as in "the woman obese". *Ambiguous errors* are those that could be classified equally well as either developmental or interlingual as in "I no have a house". Finally, *other errors* are those that do not belong to any of the previous categories as in "He do Spanish".

Therefore, the main objective of this study will be to classify the errors made by a number of Spanish students doing an engineering degree in a public Polytechnic University over the last two years, according to the comparative taxonomy of Dulay, Burt, and Krashen. We will attempt not only to categorise the errors following the aforementioned taxonomy, but also to demonstrate that the most frequent type of errors detected, at least in the case of this specific group of technical students whose English level is not the same as those students doing the English Studies degree, is not the one expected and predicted by some researchers.

LITERATURE REVIEW

Errors are recognised as those parts of conversation (spoken discourse) or composition (written discourse) that deviate from some selected norms of mature language performance. Teachers have long realised that making errors is an inevitable part of learning. According to Dulay, Burt, and Krashen, learners cannot learn a foreign language without first committing errors.

To tackle the problems occurring in the production of non-native speakers is to have a myriad on the types of errors that learners usually commit. Analysing errors as a process is a very important procedure since it provides teachers and instructors of second language with valuable information on the learner's language, helping them to improve their teaching methods (Burt & Kiparsky, 1972; Dulay & Burt, 1972; Richards, 1974).

Error analysis made a distinction between error and mistake. The first attempt to distinguish error from mistake was done by Corder (1973). He pointed out that errors reflect gaps in the learner's target language. Corder (1974:24) stated that "errors of performance (mistakes) will be characteristically unsystematic and errors of competence, systematic". As a result, the learner is unable to correct it by himself (self-correction). While mistakes are usually slips and lapses in performance. The learner knows the right form, but he commits mistakes because of nervousness, tiredness or fatigue. So, the learner is able to correct the mistake by himself.

Error analysis is the study of kind and quantity of error that occurs, particularly in the field of applied linguistics. These errors can be divided into three sub-categories: overgeneralization, incomplete rule application, and the hypothesizing of false concepts, reflected a learner's competence at a certain stage and thereby differed from learner to learner (Rustipa, 2011). From the previous statement, it can be concluded that error analysis is a theory developed by the researchers to dig and acknowledge the learners' problems in learning a second language.

According to Brown (2000:165) a mistake is performance error, "that is either a random guess or a slip in that. It is a failure to utilize a known system correctly". This means that a mistake is common between the learner of the second language and the native speaker, but a native speaker has an ability to recognize his mistakes.

Edge (in James 1998) calls all deviant forms of second language which a learner of second language produces, mistakes, and he categorizes mistakes into three groups:

- Slips appear as a result of processing problems or carelessness. According to Edge, the learner is able to correct mistakes by himself when he has an opportunity to do.

- Errors defined by Edge as "wrong forms that the pupil could not correct even if their wrongness were to be pointed out", but it indicates what the learner wanted to say.
- Attempts, Edge's last group, are "almost incomprehensible, and the learner obviously has no idea how to use the right form" (James 1998, p. 81). In this condition, learners try to use alternative communication strategy.

Ellis (2003) declared that there are two main reasons that motivate the researchers of second language to study learner's errors. Firstly, they inform us why learners commit the errors. Secondly, knowing the errors that learners commit may help teachers to enhance students' ability of self-correction. Studying any deviation helps the instructor to create remedial action in the process of errors correction (Corder, 1981). In addition, studying errors provide teachers information about the problematic areas where the students commit the most errors (Dulay, Burt & Krashan, 1982).

Following Bartholomae (1984), students inevitably make errors in their writing productions. And given the prestige that nowadays society gives to correctness in formal writing, teachers quite often correct every error they found in their students' papers or assignments. But still a wealth of scholarly research not only suggested that this approach to dealing with language error is quite ineffective, but it also argued that it may have a negative impact on their writing ability, usually because it destroys fluency.

Subsequently, error analysis was predominantly used to help language teachers predict what problems a language learner would have due to the linguistic differences between the learner's native language and the target language. In this way, errors that could not be directly attributed to language interference were virtually ignored whilst those most frequently focused upon were so well-known that many teachers found the work of these researches redundant.

Also, in the decade of the 1970s, several studies were conducted on the speech and writing of adults learning a second language. And these studies found that most of the non-phonological errors that an adult learner made did not reflect their mother tongue. However, the proportion of interlingual errors observed was larger than those which were detected in the case of children. We can include at this stage the studies developed by scholars such as Bertaku (1974), Lo Coco (1976), White (1977), Hanania and Gradman (1977), among others.

Dulay, Burt, and Krashen limited their discussion to the descriptive aspects of error taxonomies in the belief that the accurate description of errors is a separate activity from the task of inferring the sources of these errors. Therefore, they focused on error taxonomies that classify errors according to some observable features of the error itself, without any reference to its underlying cause or source.

It is precisely due to those observable features of errors that we decided to base this research on one of the taxonomies proposed by these researchers. They have called these descriptive taxonomies, reviewing the existing previous literature in order to present the most useful bases for the descriptive classification of errors. And they classified errors into the following categories: linguistic category, surface strategy taxonomy, comparative analysis taxonomy, and communicative effect. These linguistic taxonomies classified errors according to the language component and the particular linguistic constituent that the error affects.

In the first taxonomy, *linguistic category*, James also stated that errors were regarded with respect to where the error was placed in the whole system of the foreign language based on the linguistic item that is disturbed by such error. This taxonomy thus specifies in which part of language the error is located. Therefore, errors following this taxonomy can be classified according to phonology (pronunciation), syntax, morphology (clause, noun phrase, verb phrase, auxiliaries, prepositions, adjectives…), semantics (meaning), lexicon (vocabulary), and discourse (style). According to Dulay, Burt, and Krashen "constituents include the elements that comprise each language component" (Dulay, Burt, and Krashen, 1982: 146), although they recognised that a full presentation of the language components and constituents would require a summary of descriptive linguistics beyond the scope of their research.

Burt and Kiparsky developed another linguistic category taxonomy into which they classified several thousand English errors made by students learning English in foreign as well as host environments (1972). The classifications are as follows: The Skeleton of English Clauses; The Auxiliary System; Passive Sentences; Temporal Conjunctions; Sentential Complements; Psychological Predicates.

In the second taxonomy, *surface strategy*, errors are based on the diverse means in which the students' mistaken account is different from the supposed foreign language version. This means that this taxonomy underlines the ways in which the surface structures diverge. According to this taxonomy, errors are consequently divided into the following four types: omission (omit necessary items), addition (add unnecessary items), misformation (misform items) or misordering (misorder items). The classification put forward by this taxonomy gives us an accurate explanation about the cognitive processes that cause the students' reconstruction of the new language that is being learnt. As Dulay, Burt, and Krashen affirmed "errors are not the result of laziness or sloppy thinking, but of the learner's use of interim principles to produce a new language" (Dulay, Burt, and Krashen, 1982: 150).

This category highlights the way surface structures are altered. Analyzing errors from this taxonomy holds much promise for researchers as it is concerned with identifying cognitive processes that underlie the learners' reconstruction of the new language. This also gives us consideration that students' errors are not the result of laziness or sloppy thinking but rather are based on some logics, as the result of the learners' use of interim principles in order to produce a new language.

It is precisely the third taxonomy, *comparative analysis taxonomy*, the one that we are actually following so as to carry out this study, and the one we will devote more lines to explain in detail. As stated by Dulay, Burt, and Krashen, "the classification of errors in a comparative taxonomy is based on comparisons between the structure of L2 errors and certain other types of constructions" (Dulay, Burt, and Krashen, 1982, p. 163). And such comparisons have brought about four main error categories: interlingual, developmental, ambiguous, and other errors.

Researchers have consistently found that the great majority of errors in the language output of L2 learners, in our case undergraduate engineering students, is of the developmental type. Although adult learners tend to exhibit more mother-tongue influence in their errors than children normally do, adult interlingual errors also occur in relatively small numbers according to these researchers. However, as we will be determining along the present research, the great majority of the errors detected in the 72 essays examined in our study correspond to the interlingual type.

According to the researchers, "*developmental errors* are errors which are similar to those made by children learning the target language as their first language" (Dulay, Burt, and Krashen, 1982, p. 165). For example, if we take the following utterance made by a Spanish child learning English "dog eat it", the omission of both the definite article and the past tense marker may be classified as developmental because they are also found in the speech of children learning English as their first language. When such errors are made by second language learners, it would be reasonable to hypothesise that mental mechanisms underlying general language development come into play, not the rules and structures of the learner's native language.

Quoting these researchers, "*interlingual errors* are similar in structure to a semantically equivalent phrase or sentence in the learner's native language" (Dulay, Burt, and Krashen, 1982, p. 171). For example, the utterance "the man thin" produced by a Spanish speaker reflects the word order of Spanish adjectival phrases [article + noun + adjective]. In order to identify an interlingual error, what researchers normally do is to translate the grammatical form of the learner's phrase or sentence into the learner's first language so as to see if there exist similarities. According to Ellis (1985), this type of error is interference or transfer, which means that the learner's native language interferes with the learning of the L2, or rather it transfers into the learner's developing L2 system.

They chose the term "interlingual" instead of the equally similar labels "interference" or "transfer" because it seemed to be the least explanatory in connotation. Interlingual errors simply refer to L2 errors that reflect native language structure, regardless of the internal processes or external conditions that spawned them.

Ambiguous errors could be equally classified as developmental or interlingual errors. According to Dulay, Burt, and Krashen "this type of error reflects both the learner's L1 and also the type of error in the speech of children acquiring English as their first language" (Dulay, Burt, and Krashen, 1982, p. 172). For example, in the utterance "I no have a bike", the negative construction reflects the learner's native Spanish and it is also characteristic of the speech of children learning English as their first language.

The ambiguous category is particularly important in a comparative taxonomy. Assigning such errors to a distinct category guarantees the clarity of the findings resulting from a comparative analysis of errors and allows researchers to draw clear theoretical inferences from the rest of the data obtained.

Other errors make up a sort of grab bag for items that do not fit into any other previous category. For example, if we take the utterance "She do thirsty", the speaker used neither the native Spanish structure (in the use of have for is as in She have hungry), nor an L2 developmental form (such as She hungry), where the auxiliary is omitted altogether.

Such an error would go into the other category, also called "unique errors" by Dulay and Burt (1974), referring to their being unique to L2 learners. These errors are not similar to those made by children during the first language development, so they must be unique to second language learners, and since they are not interlingual, at least some of them must be unique reflections of a creative construction.

The fourth and last taxonomy is the *communicative effect taxonomy*, "based on the perspective of their effect on the listener and the reader" (Dulay, Burt, and Krashen, 1982, p. 189). It deals with discriminating between errors that seem to cause miscommunication and those that do not. Errors that distress the overall organization of the sentence usually hinder communication, while those errors that affect a single element of a sentence usually do not hinder communication. And following Dulay, Burt, and Krashen (1982), this type of taxonomy classifies errors into global errors and local errors.

The communicative effect classification deals with errors from the perspective of the effect on the listener or reader. This taxonomy focuses on distinguishing between errors that seem to cause miscommunication (global error) and those that do not (local error). A global error is one which involves the overall structure of a sentence and a local error is one which affects a particular constituent.

Global errors are errors that affect the complete sentence organization and they meaningfully hamper communication. These errors include the wrong order of major constituents as "English language use many people"; missing, wrong or misplaced sentence connectors as "He will be rich until he marry"; missing cues to sign obligatory exceptions to pervasive syntactic rules as "The student's proposal looked into the principal"; and regularization of pervasive syntactic rules to exceptions as "We amused that movie very much".

On the other hand, *local errors* are those that affect single elements in a sentence and that do not usually hinder communication significantly, at least not to a great degree. This type of errors includes errors in noun and verb inflections, articles, auxiliaries, and the formation of quantifiers, among some others.

We cannot end up this review without making allusion to several studies which were carried out by some Spanish university professors such as Roca de Larios, Murphy, and Manchon Ruiz among others. In the last decades, this group of professors have published a number of articles dealing not only with the analysis of errors but also with other aspects regarding their interpretation, strategies and components.

METHOD

The present research was conducted using a quantitative research design that was used to disclose what type of errors and how many these undergraduate students made. Not only did we attempt to classify all the errors encountered, but also, and more specifically, we did find out what the most frequent ones were. Let's remember at this point that we were following the comparative taxonomy proposed by Dulay, Burt, and Krashen in "Language Two" (1982), as we forestalled in the previous section.

The group of students included in this study encompassed an uneven number of undergraduate technical engineering students who were doing their degree on a public Spanish Polytechnic University. These students, whom were mostly male given the technical nature of the degree, were asked to carry out a written assignment during the last two academic years in order to get a bigger quantity of participants.

This uneven number of contributors was justified by shedding some light on the kind of assignment being asked for. Since the task was not compulsory but optional, not the same number of students handed in the task. Therefore, a total of 72 essays were delivered for the task although the overall distribution was slightly unbalanced in each year.

The focus of the study was thus laid on the classification of the many errors committed by these undergraduate students along the written assignment. And in order to do so we kept the comparative taxonomy classification offered by Dulay, Burt, and Krashen which divided errors into four different categories: ambiguous, developmental, interlingual, and other errors. For the sake of clarity, we should clarify that these categories have been arranged here alphabetically for the moment.

Consequently, the instruments which were used to develop this current research referred to the writing assignment that these students had to undertake along the semester. At this point the author figures out that it is important to remark an important fact about the assignment under scrutiny. Although the students had to cope with writing an essay about a specific topic as we will later detail, it is also true that prior to this writing they received the necessary background information in terms not only of grammar of vocabulary but also of the required structures, which would apparently enable them to deal with the assignment appropriately. The following were the steps in analyzing the data:

1. Classifying the erroneous clauses using the comparative analysis taxonomy by Dulay et al, (1982) by putting them in the following tables in order to make the process of analysis easier.
2. Providing the correct clauses that should be used in the texts as the correction of the erroneous clauses.
3. Drawing conclusions. Once the errors were calculated and arranged, the writer made the result of the analysis in the form of percentage.

Finally, regarding the procedure followed, this study was conducted during one semester in the last two consecutive academic years. The collection of data did not take place until all the necessary information had been studied and dealt with in class, on the one hand, and they also had enough time to carry out the assignment, on the other. Although many different errors were encountered, only those falling under the comparative taxonomy were taken into consideration.

RESULTS

Behind the analysis of the data which we obtained, in general terms, we can assert that a total of 256 errors were encountered along the 72 essays examined, which gives us a rough average of 3.5 errors per essay. Taking into account that there was a number of essays with no errors at all, this figure gives us an accurate idea on the number of errors which were detected along the task.

And following the comparative taxonomy of errors proposed by Dulay, Burt, and Krashen in 1982, the errors found in the task were classified into the following types: ambiguous errors, developmental errors, interlingual errors, and other errors. Errors that did not fall under these categories were not considered in this research.

The task that students had to accomplish along the semester was the description of the electricity cycle (generation, transmission, and distribution) by means of including action verbs and the use of the passive voice. This task was carried out after the students had completed a unit of the syllabus on electricity in which the electricity cycle was discussed.

The number of essays underwent for this task was 72 in total. And the distribution of errors was quite odd. If the most commonly found essays in this task were those with 5 or more errors, with a subtotal of 25 cases representing a 34.8%, the least ones were those with 1 or 2 errors with 13 cases, standing for an 18% overall. In between we came across 19 essays with no errors, denoting a 26.4 of the total, and 15 essays with 3 or 4 errors, meaning a 20.8% as a whole. This allotment gives us a clear idea of how distributed the number of errors detected were in this initial task.

Should we focus on the 53 compositions containing 1 error at least, we came across the fact that the category with the highest number of errors in this task following the comparative taxonomy was interlingual, with a total of 99 errors out of 256, which represented a 38.8% of the total. In the second place we found developmental errors with 61 cases detected, representing a 23.8% overall. Other errors came in the third position with 59 errors spotted, standing for a 23% of the total. And in the last place we found ambiguous errors with 37 examples which meant a 14.4% in general. The numbers and percentage of each error category are shown in the following table:

Table 1. Number and percentage of error categories in the task.

Type of Error	Number of Errors	Percentage
Interlingual errors	99	38.8%
Developmental errors	61	23.8%
Other errors	59	23%
Ambiguous errors	37	14.4%
TOTAL	**256**	

Source: Own elaboration

Let's now have a look at *interlingual errors*, the category with the highest number of errors, 99 out of 256. This meant that almost one in each three errors belonged to this category in the task. And as it will be shown in table 2, interlingual errors were divided into the following six types keeping the division made by Krashen, Burt, and Dulay: addition of past tense, addition of preposition, verb-number disagreement, inappropriate demonstrative, inappropriate preposition, and inappropriate pronoun.

Table 2. Number and percentage of interlingual errors in the task.

Type of Error	Number of Errors	Percentage
Addition of past tense	1	1%
Addition of preposition	1	1%
Verb-number disagreement	74	74.8%
Inappropriate demonstrative	6	6%
Inappropriate preposition	12	12.1%
Inappropriate pronoun	5	5%
TOTAL	**99**	**38.8%**

Source: Own elaboration

Under the category of interlingual errors, the type with the highest number of cases in the students' compositions was that of verb-number disagreement with 74 errors detected, which represented a 74.8% of the total. Given the technical nature of the task, students seemed to be unable to maintain the agreement between the subject and the verb as in the following example: "The smart grid is a grid that <u>use</u> information technology". This is a clear influence of their mother tongue, especially when there is a deviation between the subject and the verb to concur.

Quite less numerous was the next type of interlingual errors, the use of an inappropriate preposition, something which happened 12 times, representing a 12.1% overall. The following sentence is an example of this type of error: "The process of electricity is controlled <u>for</u> electronic and computing means". Again, we can find a direct influence of the mother tongue since the mistaken prepositions were due to a direct translation into English.

Then we found 2 other types of interlingual errors close in numbers. On the one hand, there was the use of an inappropriate demonstrative, with 6 errors detected which made up for a 6% of the total. Using the correct demonstrative in the sentences did not represent a big deal for these students. An example of this type of error could be the following: "In <u>this</u> transformers the voltage is reduced by electromagnetic induction". Incorrect associations by the students were done in the use of demonstratives, without reflecting on the number differences.

And on the other hand, there was also the use of an inappropriate pronoun with 5 cases spotted, meaning a 5% overall. As can be seen both types shared closed number of errors. The following could be included as an example of such an error: "When electricity reaches <u>her</u> destination, it is stepped down to a low voltage in a substation". The use of impersonal pronouns was not recognised by the students, tending to overgeneralise it by their male or female correspondents.

Finally, the presence of two other types of interlingual errors, the addition of a past tense and the addition of a preposition, was rather testimonial in this first task. In each type there was just one error, which signified a scarce 1% of the total. The following can be included as an example of the former type, the addition of a past tense: "We can classified into 2 components". And the next sentence is an example of the latter type, the addition of a preposition when it was not needed: "The electric power which is generated is stepped up to a higher voltage at which it connects to the electric power transmission network". These residual cases do not have an overall applicability to the entire group so we cannot establish a proper justification which can validate their occurrence.

The second most numerous category in this task corresponded to *developmental errors*, with a total of 61 errors out of 256, which represented a 23.8% overall. This time developmental errors were classified into the following eight different types: omission of auxiliary, omission of plural marker, omission of regular past tense marker, omission of irregular past tense marker, addition of regular past, addition of irregular past, inappropriate part of speech, and omission of the infinitive marker "to". The distribution of each type of developmental error can be checked in the next table.

Table 3. Number and percentage of developmental errors in the task

Type of Error	Number of Errors	Percentage
Omission of auxiliary	12	19.7%
Omission of plural marker	12	19.7%
Omission of regular past tense marker	15	24.6%
Omission of irregular past tense marker	1	1.6%
Addition of regular past	1	1.6%
Addition of irregular past	2	3.3%
Inappropriate part of speech	17	27.9%
Omission of infinitive marker "to"	1	1.6%
TOTAL	**61**	**23.8%**

Source: Own elaboration

In this task, there was a relatively even distribution in the number of errors not only for the most recurrent errors but also for the least frequent ones as can be deduced from the information gathered in the preceding table. Whether we came across four different types of developmental errors in the upper side of the rank, the same happened in the lower side with four other types of errors.

Should we just start with the most numerous type of developmental error, this referred to the misuse of an inappropriate part of speech, error which occurred 17 times, representing an almost 28% of the total of developmental errors. Nearly one third of developmental errors belonged to this type. Apparently, students faced some problems when they had to construct long complex sentences keeping the logical and normal order. An example of this type of error could be the following: "This generation is far from consumption and so is produced in the central electricity". Building long technical sentences usually posed some problems in order to follow a natural order in English.

Close in frequency we could find the next type of developmental error, the case of the omission of a regular past tense marker, something which happened 15 times along the task, which stood out for a 24.6% overall. They seemed to forget to include the past tense marker in some verbs as can be seen in the following example: "The substation is where the electricity voltage is increase through the transformers". It was quite surprising to come across this type of error and, above all, its frequency when they should have already acquired it.

Two other types of developmental errors followed in number in the upper side of the rank: the omission of the auxiliary, and the omission of a plural marker. In each case, we found 12 errors in the task, denoting a 19.7% of the total. The following can be included as an example of the former type, the omission of the auxiliary: "Why __ we do it?". And the next one is an example of the latter case, the omission of a plural marker: "The electrical grid is one of the most important invention in the modern world". Using the auxiliary when forming questions and keeping the concordance in plural formation seemingly appeared to pose some problems when writing this task. The former reflects the influence of the mother tongue since the use of auxiliaries is not necessary in Spanish.

On the other hand of the scale we came across four different types of developmental errors whose occurrence along the task was nothing but testimonial with just 1 or 2 errors. One of these types was the addition of an irregular past formation in the sentence with 2 errors spotted, which represented a 3.3% of the total. This sentence can serve as an example for such a type of error: "We will lost a lot of its energy".

The other three types of developmental errors, which merely happened once in each case, standing for a 1.6% of the total, included the omission of an irregular past tense marker, the addition of a regular past, and the omission of the infinitive marker "to". This symbolic presence in the task denote that these types of errors did not suppose a big problem to accomplish.

The following sentence could be considered as an example of one of these types, the omission of an irregular past tense marker: "This electricity is then send to a step-up transformer". As an example of the addition of a regular past tense form we could include the following: "When we use step-up transformers, we will decreased the intensity of the electricity". Finally, the next one could be cited as an example of the omission of the infinitive marker "to": "When electricity is used, it's necessary __ decrease it". Again, its scarce occurrence prevented us from drawing definite conclusions on their misuse.

In the third place in terms of number of errors, there came the category of *other errors* with 59 errors encountered out of 256, which embodied a 23% of the total. However, it should also be clarified that this figure is a bit deceitful since a great deal of these corresponded to the case of incorrect spelling. Taking into account the technical nature of the task and the specific vocabulary they were asked to use, it may seem reasonable to expect such a number of words incorrectly written.

Other than this, the presence of other errors in this task was quite symbolic as we will now explain. For this task, other errors were classified into the following five different types: addition of article, addition of plural marker, omission of progressive form, incorrect spelling, and addition of subject.

Table 4. Number and percentage of other errors in the task.

Type of Error	Number of Errors	Percentage
Addition of article	2	3.4%
Addition of plural marker	3	5.1%
Omission of progressive form	5	8.4%
Incorrect spelling	46	78%
Addition of subject	3	5.1%
TOTAL	**59**	**23%**

Source: Own elaboration

As anticipated, the bulk of other errors corresponded to the case of incorrect spelling with 46 errors detected, denoting a 78% of the total. Given the scientific nature of the task, we may expect such a number of words incorrectly written. A clear example of this type of other errors can be the following: "Industrial consummers are large facilities like manufacturing plants or office building". The difference in frequency between this type of errors and the rest of other errors was quite outstanding.

Therefore, it was from a notably distance that we encountered the next type of other errors, the omission of a progressive form. This type happened 5 times, representing an 8.4% overall. The omission of the progressive form of the verb was reflected in cases like the following: "The electrical grid is the network which is generate power". Perhaps different tenses were mixed up by the students in their attempt to construct the sentences.

Next, we found two other types of other errors with the same number of incidences, 3, standing for a 5.1% as a whole: the addition of a plural marker and the addition of a subject. The former case, the addition of the plural marker, can be seen in the following example: "The generation of electricity can be in differents plants like hydraulic or nuclear". This clearly reflects the influence of the mother tongue adding number also to adjectives as it usually happens in Spanish. And the following could be cited as an example of the latter, the addition of a subject when it was not required: "As a consumer, you can't choose which it is your distribution company".

The last type of other errors corresponded to the case of the addition to the sentence of an article when this was not needed, something which just happened twice, representing a 3.4% of the total. An example of this type of error could be the following: "Nowadays, the most part of the machines are fed by electricity". This is a neat case of direct translation from Spanish since we use the article in the given construction.

Finally, the least numerous category of errors in this first task following the comparative taxonomy corresponded to *ambiguous errors*, with a total of 37 errors out of 256, representing thus a 14.4% of the total. In this initial task, ambiguous errors were divided into the following types: omission of the article, omission of the subject, inappropriate quantifier, and inappropriate article.

Table 5. Number and percentage of ambiguous errors in the task

Type of Error	Number of Errors	Percentage
Omission of article	17	46%
Omission of subject	14	37.8%
Inappropriate quantifier	3	8.1%
Inappropriate article	3	8.1%
TOTAL	**37**	**14.4%**

Source: Own elaboration

As can be seen in the previous table, there is a certain degree of parallelism with the category of developmental errors in the number of cases found both in the upper side of the scale and in the bottom. But this time the frequency is almost reduced to the half.

The most frequent type of ambiguous errors was the case of the article, either definite or indefinite, being incorrectly omitted in a sentence. This occurred 17 times in the task, representing a 46% of the total as can be seen in the following example: "Electricity is generated at __ generating plant". This sole type of ambiguous errors meant almost half of the total of errors in this category. Students tended to overgeneralise the omission of the article when it was not necessary.

The next type of ambiguous errors in frequency was the case of the omission of the subject in the sentence, something which happened 14 times, denoting a 37.8% overall. As an example of this type of error we could include the following: "The electricity goes through a lot of kilometres until our homes because __ is safer for us". Including the correct subject in each case posed some problems on the students, especially when this was an impersonal one, reflecting their Spanish influence.

Finally, we found out two other types of ambiguous errors in the lower side of the rank with just 3 errors each one, standing for an 8.1% of the total: the use of an inappropriate quantifier and the use of an inappropriate article. In the case of former type, the use of an inappropriate quantifier, we could include the following sentence: "The most application of the electrical grid is the smart grid". Finally, as an example of the latter type, the use of an inappropriate article, the next one could be included: "Now there is a only electricity central for all types of sectors". Again, their low occurrence makes impossible to get definite conclusions or neat evidence which justifies their use.

DISCUSSION OF RESULTS

Writing is one of the most difficult skills for EFL students to acknowledge. Producing a well-organised piece of writing seems to be a difficult task for learners, especially if we bear in mind the notable fact that we were dealing with undergraduate students doing a technical degree and whose domain of the FL should then not be comparable to that of those students doing English studies or any related field of expertise.

Error Analysis included an internal focus on the learner's creative ability to construct a foreign language. Therefore, it was the main method used to document the systematic errors appearing in the writing compositions of EFL students. Errors were usually classified according to the language component: phonological, morphological, syntactic, etc.

Throughout the present research, errors have been arranged according to the classification proposed by Dulay, Burt, and Krashen in "Language Two" (1982) into four taxonomies or categories: linguistic category, surface strategy taxonomy, comparative taxonomy, and communicative effect taxonomy.

And more precisely, we have been following the comparative taxonomy in this study whereby errors were organised into the next four main categories: ambiguous errors, developmental errors, interlingual errors, and other errors (being arranged here alphabetically without any other restraining factor).

The first important aspect to recall back is the fact that some of the researchers mentioned in the present study defended that the majority of errors that adult FL learners committed corresponded to the developmental category. However, we have consistently corroborated that it was not the case in our research. We have definitely shown that the most frequent errors that undergraduate technical engineering students made were interlingual in the case of the written assignment examined, contrary to the opinion of these researchers.

Secondly, the data obtained in this research came from the thorough analysis of the 72 essays handed in by Spanish undergraduate engineering students during the last two academic years. These students had to deliver a task along the semester with an uneven distribution throughout the two courses.

Thirdly, concentrating on the number of errors, a total of 256 errors were detected in the written assignment with an uneven distribution. A total of 256 errors were found in the 72 essays of the task, which gave us an estimated average of 3.5 errors per essay.

In the fourth place, we can also emphasise the number of errors detected in the 72 essays. In this, we made a distinction between essays without errors, essays having one or two errors, essays having three or four errors, and essays having more than five errors. And in terms of numbers, the first group was that composed of essays having more than five errors with a total of 25 essays, representing a 34.7% of the total. Next, we found essays with no errors, 19 out of 72, standing out for a 26.4% in all. In the third place, there were essays with three or four errors, 15 out of 72, which represented a 20.8% of the total. Finally, the least representative group was that of essays with one or two errors, with 13 essays in total, denoting a 18.1% as a whole. These data can easily be interpreted in the next explanatory table.

Figure 1. Number of errors in the task
Source: Own elaboration

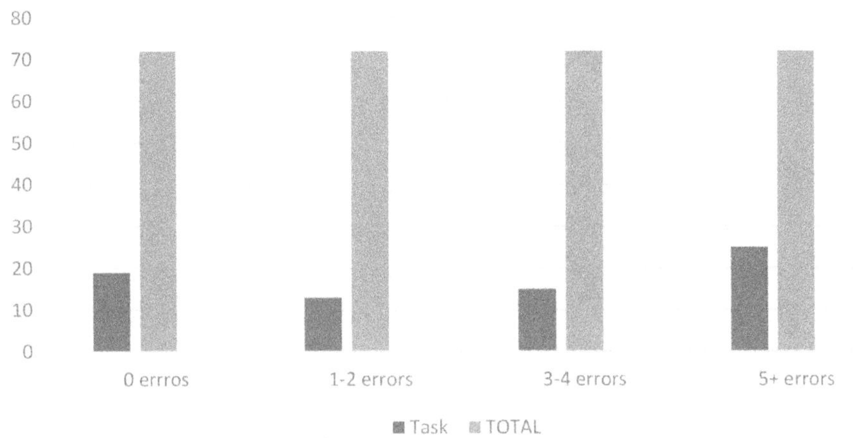

In the fifth place, we focused in depth on the different categories of errors encountered along the 72 essays. And contrary to the view of some researchers, we determined that interlingual errors, and not developmental errors as suggested, was the category with the highest number of occurrences along the task with a total of 99 errors, making up a 38.7% of the total. In second place there came developmental errors, with 61 errors in total, representing a 23.8% overall. The third place was for other errors with 59 cases detected, which denoted a 23.1% in all. Finally, the last representative category was that of ambiguous errors with 37 errors, which represented a 14.4% in general. In the next table we can appreciate the distribution of errors according to the category.

Figure 2. Type of errors in the task
Source: Own elaboration

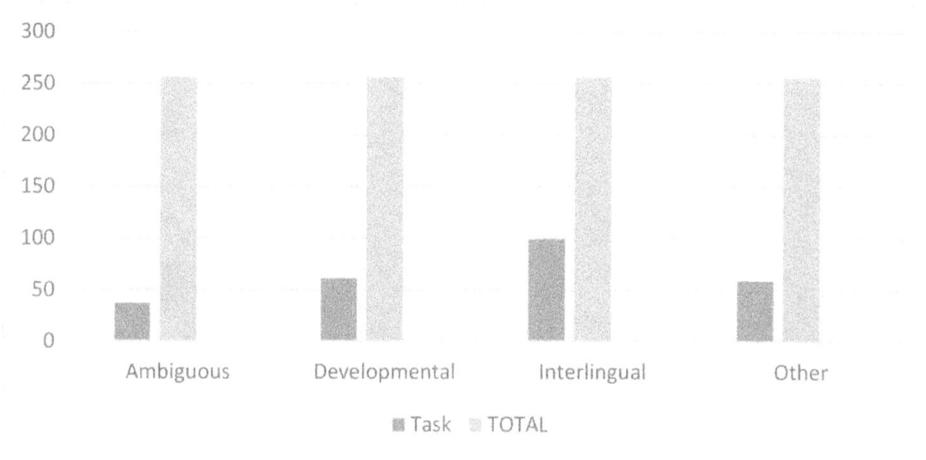

The most repeated type of error was the case of verb-number disagreement with 74 errors whilst the least repeated ones were the addition of past tense and the addition of a preposition with just one instance each. Other frequent interlingual errors were the use of an inappropriate demonstrative, inappropriate preposition or inappropriate pronoun. In the following pie chart, we can acknowledge the distribution of types of interlingual errors.

Figure 3. Distribution of types of interlingual errors
Source: Own elaboration

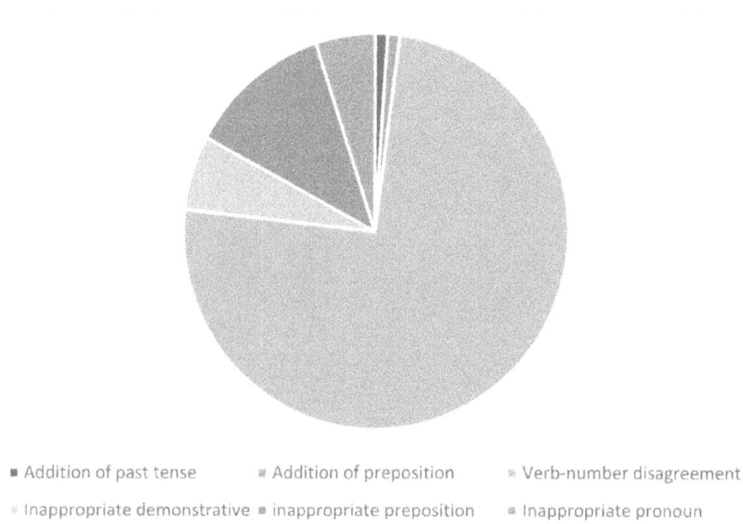

The second category corresponded to *developmental errors*, which were divided into the following eight types: omission of an auxiliary, omission of plural marker, omission of regular past tense marker, omission of irregular past tense marker, addition of regular past, addition of irregular past, inappropriate part of speech, and omission of infinitive marker “to”.

Within developmental errors, the most frequent type was the use of an inappropriate part of speech with 17 cases. On the other hand, the least frequent types were the omission of irregular past tense marker, the addition of regular past, and the omission of the infinitive marker “to” with just one case each. Other types which were also recurring were the omission of the auxiliary or the omission of a plural marker, as can be seen in the next pie chart.

Figure 4. Distribution of types of other errors
Source: Own elaboration

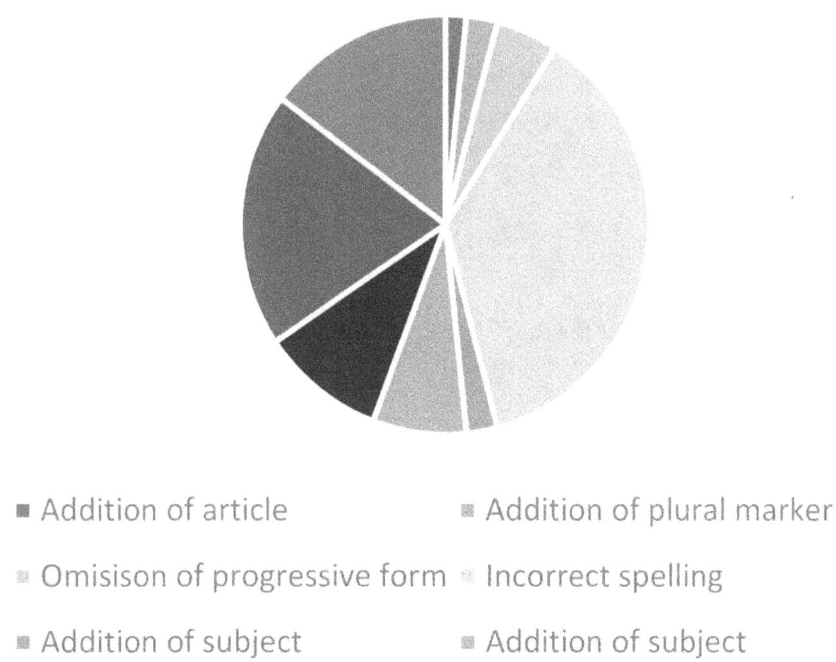

In the third place, *other errors*, which were divided into the following five types: addition of article, addition of plural marker, omission of progressive form, incorrect spelling, and addition of subject. If the most represented type was that of incorrect spelling with 45 cases, the least represented one was the addition of the article with just 1 error. Some other recurrent types were the addition of a plural marker and the omission of the progressive form.

At this point, it should be reminded that had not been for the type of incorrect spelling, the category of other errors would have come in the last place in the rank. However, it may seem unavoidable to expect such a high number of words incorrectly written in the assignments. We can appreciate it better in the following pie chart.

Figure 5. Distribution of types of other errors
Source: Own elaboration

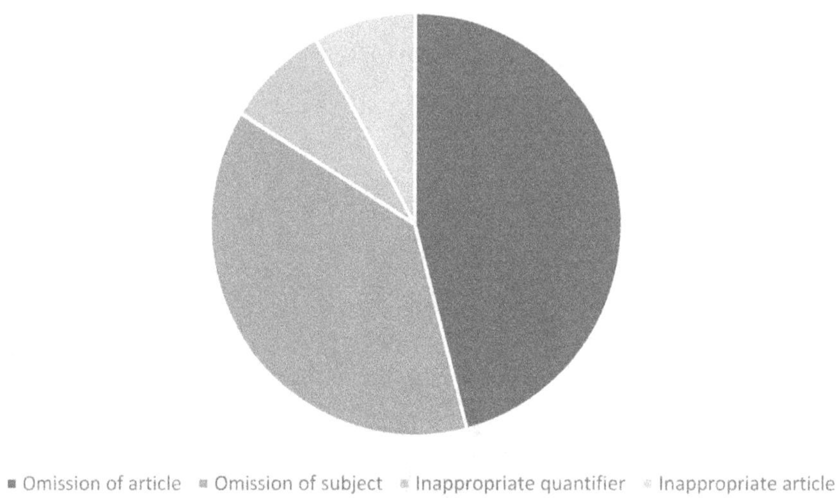

Finally, *ambiguous errors*, the last category in numbers, were divided into the following four types: omission of article, omission of subject, inappropriate quantifier, and inappropriate article. Whilst the first two types, the omission of article and the omission of the subject were found quite recurrently each, on the other hand the use of an inappropriate article and the use of an inappropriate quantifier were just encountered 3 times as can be deduced from the following pie chart.

Figure 6. Distribution of types of ambiguous errors
Source: Own elaboration

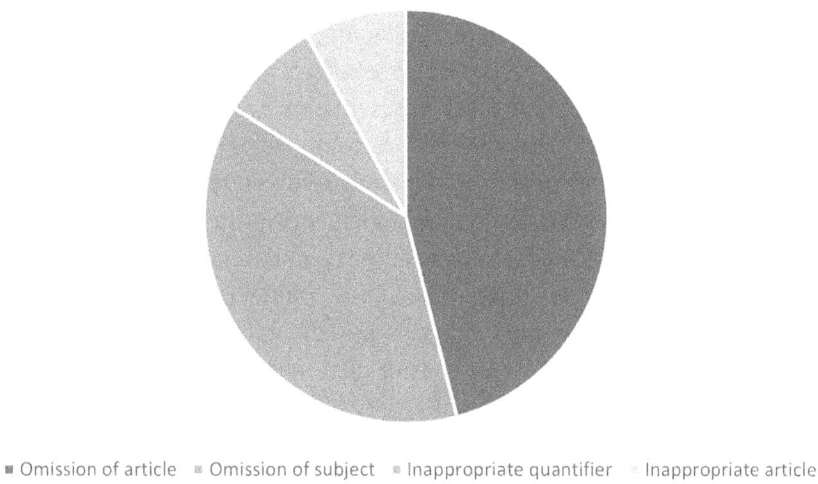

CONCLUSION

Referring to the conclusions of the study, there are some suggestions to offer. Firstly, the students should learn more about their ways of constructing sentences including the use of some grammatical features so that they can improve their writing skills by reducing their common-made errors. Secondly, the teachers are expected to make use of the results of the error analysis in order to help students improve their writing skill and reducing their errors as well as constructing a plan and emphasizing what they need to bring to the class. Lastly, since this study is limited to one single class of engineering students only, further researchers are possible to expand the number of subjects to see whether what lies beyond the students' writing errors is the same or different to the one revealed in this study.

As a closing remark, we can finish withdrawing the main information collected along the study. The corpus of this research has been the analysis of errors in 72 essays written by technical engineering students during the last two academic years in a written assignment. And in order to do so, we followed the comparative taxonomy proposed by Dulay, Burt, and Krashen. The findings in these essays have demonstrated that the most recurrent category of errors detected was that of interlingual errors, followed by other errors and developmental errors. Finally, we placed ambiguous errors in the last position in the rank.

We can also emphasise the number of errors detected in the 72 essays. In this, we made a distinction between essays without errors, essays having one or two errors, essays having three or four errors, and essays having more than five errors. And in terms of numbers, the first group was that composed of essays having more than five errors with a total of 25 essays, representing a 34.7% of the total. Next, we found essays with no errors, 19 out of 72, standing out for a 26.4% in all. In the third place, there were essays with three or four errors, 15 out of 72, which represented a 20.8% of the total. Finally, the least representative group was that of essays with one or two errors, with 13 essays in total, denoting an 18.1% as a whole.

To finish with, we would also like to include some suggestions for these students in they want to improve their Technical English writing. Firstly, they should try to read in English, which allows them to acquire vocabulary and become familiar with the different grammatical structures. Then, they could write a little every day, since daily practice is infallible to improve writing. In the third place, and it is something they usually forget about, they can consult dictionaries, taking advantage of it to look up synonyms and antonyms. Finally, they should always review what they have written, read it aloud to check for errors.

REFERENCES

Bartholomae, D. (1984). The Study of Errors. *College Composition and Communication*, 31.

Bertkau, J. (1974). An Analysis of English Learner's Speech. *Language Learning*, *24*(2).

Brown, H. D. (2000). *Principals of language learning and teaching*. Longman.

Burt, M., & Kiparsky, C. (1972). *The Gooficon: A repair manual for English*. Newbury House Publishers.

Corder, J. P. (1981). *Error Analysis and Interlanguage*. Oxford University Press.

Dulay, H., & Burt, M. (1974). Errors and Strategies in Child Second Language Acquisition. *TESOL Quarterly*, *8*(2), 8. doi:10.2307/3585536

Dulay, H., & Burt, M. (1974). Natural sequences in child second language acquisition. *Language Learning*, *24*(1), 37–53. doi:10.1111/j.1467-1770.1974.tb00234.x

Dulay, H., Burt, M., & Krashen, S. (1982). *Language Two*. Oxford University Press.

Ellis, R. (1985). *Understanding Second Language Acquisition*. Progress Press.

Ellis, R. (2003). *The Study of Second Language Acquisition*. Oxford University Press.

Hanania, E. A., & Gradman, H. L. (1977). Acquisition of English Structure. *Language Learning*, 27.

James, C. (1998). *Errors in Language Learning and Use*. Longman.

Lococo, V. (1976). A Comparison of Three Methods for the Collection of L2 data: Free Composition, Translation, and Picture Description. *Working Papers on Bilingualism*, 8.

Manchon Ruiz, R.M., Murphy, L. & Roca de Larios, J. (2007). Lexical retrieval processes and strategies in second language writing: a synthesis of empirical research. *International Journal of English Studies*, 7.

Murphy, L. & Roca de Larios, J. (2001). Some steps towards a socio-cognitive interpretation of second language composition processes. *International Journal of English Studies*, 1.

Nzama, M. V. (2010). Error Analysis: A Study of Errors Committed by Isizulu Speaking Learners of English in Selected School. *International Journal of Humanities and Social Science*, *8*, 116–125.

Richards, J.C. (1971). A Non-Contrastive Approach to Error Analysis. *Journal of ELT*, 25.

Richards, J. C. (1974). *Error Analysis: Perspective on Second Language Acquisition*. Longman.

Roca de Larios, J. (1996). Linearization strategies in ELF writing. *Lenguaje y textos*, 8.

Roca de Larios, J., Manchon Ruiz, R. M., & Murphy, L. (2007). Componentes básicos y evolutivos del proceso de formulación en la escritura de textos en lengua materna y lengua extranjera. *Revista Española de Lingüística Aplicada*, 20.

Rustipa, K. (2011). Contrastive analysis, error analysis, interlanguage and the implication to language teaching. *Ragam Jurnal Pengembangan Humaniora*, *11*(1), 16–22.

White, L. (1977). Error Analysis and Error Correction in Adult Learners of English as a Second Language. *Working Papers in Bilingualism*, 13.

Chapter 2
From Error Analysis to Effective Writing Outputs:
An EAP Process-Oriented and Social-Constructivist Perspective

Ali Kurt
https://orcid.org/0000-0001-7197-8979
Istanbul Esenyurt University, Turkey

EXECUTIVE SUMMARY

This chapter investigates the overlap between error analysis, process-oriented writing, and a social constructivist approach to teaching and their effects on foreign language technical writing. The study is significant for its unique emphasis on the coalescence of these three domains, which is natural yet rare in research. Data collection tools included interviews, document analysis, and random participant observation. Error analysis was also used as a methodological approach to examine longitudinal data from a cohort of B1-level English language learners. The initial technical writing work revealed a failure to adopt the nature, structure, and content of technical writing. Yet, later-produced works exhibited better development over time after the interventions, which were full of specific strategies. By validating the synergy in error correction between process writing, which clears clouded minds with its precise step-by-step layout, and social constructivism, which helps generate meaning and cooperate, the study concludes the necessity of such integration into technical writing classes.

DOI: 10.4018/978-1-6684-6222-5.ch002

INTRODUCTION

When learning a new language, progress might be quick or sluggish, challenging or simple, depending on a variety of factors (Leong & Ahmadi, 2017; Selinker, 1972). One of the most challenging aspects of learning a language is developing proficient writing skills, which is more of a process than a quick one-time action (Hyland, 2003; Sarwat et al., 2021). Because writing is used as the major form of evaluation from primary school through graduate studies, this skill is significantly crucial. Besides, writing, according to Berninger and Winn (2006), is an intellectual struggle that involves testing memory, thinking about ideas (cognitive), communicating them (linguistic), and acting in a confidential way (psychological), and therefore necessitates multiple steps before ideas are put down on paper (or digital platforms) in a meaningful and coherent way (Galbraith, 2012).

Undoubtedly, students in higher education must have strong writing abilities since their written work accounts for a significant portion of their overall grades, which leads us to address English in academic contexts (Sağlamel & Aydoğdu, 2021). English for academic purposes (EAP) emerged from the larger field of English for specific purposes (ESP) and is characterized by its emphasis on teaching English to support those studying or researching in English, primarily in higher education (Hyland & Hamp-Lyons, 2002). EAP and ESP may be distinguished from other studies in English by their strong emphasis on the need to recognize and resolve student concerns related to academic work (Rao, 2020). Among the four macro-skills within the scope of EAP that students need is perhaps writing, given the faculty's high demand on students as part of the learning and assessment process (Flowerdew, 2016). To convey messages within this challenging academic context, foreign language (L2) learners should take into consideration content, organization, originality, fluency, accuracy, and the utilization of discursive rhetorical forms and style, which takes the lead over other components of writing (Richards & Renandya, 2002; Saito & Hanzawa, 2018). Thus, any help in the improvement of technical writing tasks is much appreciated given the difficulties in comprehending the produced works and creating effective outputs. Technical writing is getting more and more important to businesses, and skilled writers are in high demand (Priya & Divya, 2016). A company that keeps useful documentation sees more sales, more trust from customers, more value added to the product, and less time spent on product support. Without clear, precise writing that clearly sets out the product's specifications and how to use it, it can become, at best, useless and, at worst, dangerous. Thus, technical writing courses should deliver products and processes as a frame of reference (Larissa & Valeria, 2019).

BACKGROUND

There are essentially four distinct writing styles: expository, narrative, persuasive, and descriptive. Technical writing falls under expository writing, which is an attempt to explain a concept by separating the author's opinions from facts, statistics, cause-and-effect relationships, and evidence related to the topic (Karney & Filion, 2000). Such factual-based technical writing also entails presenting data in a logical and appropriate order (Eunson, 2011) and a neutral manner, and thus is prescriptive and written in the 'third person' language. From a different angle, technical writing is a factual-based prescriptive style in which the author presents how something is done, made, or does function, leaving out his/her opinions (Brown, 2001). Should an L2 learner communicate messages in this style, L2 learners must master the specialized vocabulary of their discipline or field of study. Britton (1965) defines technical writing as the process of gathering information from experts and sources and then composing it in a clear and comprehensible manner. However, technical writing, as Hyland (2003) points out, comes with significant challenges. First, psychological handicaps are attributed in part to students' lack of confidence, which may originate from a perceived deficiency in the skill and unfavorable attitudes toward writing. Secondly, because of their limited grammatical and lexical competence, a lack of practice with writing tasks, and inadequate strategies, L2 students may struggle with linguistic problems in writing (Peter & Singaravelu, 2021). Pedagogical challenges, on the other hand, can be traced back to teachers' weak teaching skills that might impede rather than ideally promote students' capacity to comprehend and carry out assigned writing tasks (Gill & Jangua, 2020). Cognitive impairments, likewise, impose a burden on L2 authors in idea generation, sequencing, and presenting them in an efficient and coherent style (Rana, 2018).

The premise that even native English speakers make mistakes in their writing, especially when faced with the complex demands of technical writing (Annett, 1982) provides some justification for the clumsiness that is unavoidable in the work of EFL students. Every grammatical error weakens the overall impact of the paper, so it is important for both L2 authors and their teachers to pinpoint what causes these mistakes in the first place. This is called Error analysis, a process-oriented method that helps L2 learners recognize, evaluate, and classify the deviations that occur between their language use and the norms of the target language. The analysis can be done by the learner themselves or by a trained professional (Brown, 2001). On the other hand, James (1998) views Error Analysis as the process of analyzing learners' errors by comparing what they have learned to what they lack. It also involves explaining the faults so that they may be minimized precisely. As the first scholar to bring the topic of error analysis to light, Corder classified such analysis under two types: theoretical and applied. The former seeks answers to how L2 learners acquire a foreign language, whereas the latter endeavors to enhance students' linguistic competence more rapidly and efficiently for educational ends (Corder, 1967). No matter what type, the end goal of teachers is to use error analysis in the best way possible to provide students with more accurate and engaging lessons that minimize their errors with less time wasted.

Process-oriented technical writing is a technique that focuses on the process rather than the end product, allowing L2 learners to understand how to work through and discover strategies related to how authors typically create (Brown, 2001; Zhou & Wei, 2018). Students' individual needs, goals, skills, and talents are prioritized in the process-oriented approach to writing instruction. Process-writing emerged and is now widely recognized for two prominent reasons, the first of which is researchers' and teachers' recognition that students perform better when they are immersed in an encouraging and motivating learning environment.(Reid, 2001). Discontent with a strictly regimented class ambiance and the conventional point of view in writing classes have also contributed to the trend toward process-oriented writing pedagogy (Kroll, 1990). Secondly, in contrast to product-oriented writing, the effectiveness of the process approach stems from the fact that it trains L2 writers to be more effective, competent, and keen (Nunan, 1991). Process orientation encourages learners to regulate their own learning and, in turn, their writing. This autonomy unquestionably sparks students' own intrinsic motivation, which in turn aids the development of their written language skills. In this natural progression, Murray (1972) stated process-oriented writing is an effective teaching approach for helping students understand and broke down the writing process basically into the stages of pre-writing, writing, and revision. However, a contemporary process-oriented writing consists of five stages: *planning*, in which students determine what and how to write; *drafting*, where the first draft is written; *editing*, during which individuals reflect on and *revise* the paper; and *final revision*, where the final copy is written for readers.

The prospective role of social constructivism in technical writing is based on the realization that the process of developing sufficient competence in writing calls for not only individual effort but also collaborative work with peers and teachers. In the view of constructivism (Vygotsky, 1978), learning is an active process by which learners actively create meaning out of knowledge, unlike a passive orientation in which they are merely handed information. They build meaning in a way that is unique to them, drawing on their own experiences. Vygotsky expanded on this view by arguing that human development occurs in a social context, where we all learn from and alongside one another while socially constructing meanings.

In a conventional social-constructivist methodology to assist acquiring L2, once the initial draft is completed, it is refined with the support of peer and teacher conferencing (Schreiber & Valle, 2013) as learning is not confined solely to the individual (Vygotsky, 1978). Students can harness the natural energy of other assisting sources in this case to encourage a thorough gasp and critical synthesis of target content, allowing them to actively construct and absorb their own meaning of the concepts (Powell & Kalina, 2009). With its great emphasis on the effect of social and cultural factors on learners' progress in L2 writing, such approach underpins instructional activities and intervention by instructors (Wilson & Deverux, 2014), as the more knowledgeable other (Vygotsky, 1978). Students' social engagement and participation, as a result of getting more involved in cooperative efforts with instructors and peers, reflect on their growth both inside and beyond the classroom (Nurfaidah, 2018). How, therefore, do technical writing and social constructivism go hand in hand to assist us in cracking technical writing tasks? This chapter discusses the outcomes of this interaction.

After briefly outlining the key notions upon which the study rests, it is necessary to describe another crucial concept: *interlanguage*. It is a process involving skill in a L2 and native language (L1) contrastive interference, and it reflects the misuse of a target language due to the influence of the learner's L1. In other words, in technical writing, interlanguage is the interference of L1, which is the primary impediment since it may make more challenging the process of organization of ideas, register, and text arrangement along with advanced grammatical and lexical competence. Here, mistakes are systematic and lead learners to jump to conclusions or plausible hypotheses that might result in overgeneralization and misapplication of newly learned norms (Selinker, 1992). When L2 learners are provided with correct feedback on their technical writing attempts, they are more likely to amend these mistakes and reach sufficient proficiency over time with the right exposure and practice. This constant effort toward accuracy builds up and increases the quality and complexity of writing assignments, which develops the zone of proximal development of learners, thus interlanguage (Vygotsky, 1986). It is important to remember that, as the learner has not yet mastered L2, with its own set of rules, principles, and regulations, interlanguage, thus making mistakes, is a natural process of foreign language acquisition, just as it is when kids acquire their mother tongue. Regarding the distinctive features between error and mistake so that no misunderstandings arise, Brown (2000) notes the difference between the two stances by arguing that a mistake can be made by a student who accidentally uses the language incorrectly despite having learned sufficient competence to avoid doing so. In this circumstance, the student is self-reflective enough to identify the error and devise a solution without assistance from an outside source. An error, on the other hand, is a repeated mistake committed by a student who is unable to identify the so-called faults in the target language; in other words, a student who is uninformed that what s/he says or writes is incorrect. Errors are indicators of incomplete and faulty work, which may be caused by fatigue, incompetence, and a lack of attention, as well as the adverse influence of the native language on L2 (Touchie, 1986). This is the point at which a facilitator or an outsider should step in and remedy the student's error. Through such feedback and mastering the language over time, L2 learners become better versions of themselves. A synopsis of the study has been briefly presented to enhance our understanding of the concepts under inquiry until now. The next section reflects the details of the study's background and results.

MAIN FOCUS OF THE CHAPTER

Main Body

The purpose of this chapter is to examine whether error analysis designed around the process-oriented writing classes within a social-constructivist approach may help L2 learners enhance their technical writing proficiency. Technical writing is a challenging task for L2 authors and is considered not for everyone as it requires paying close attention to factual information, clarity, accuracy, well-designed structure, specific genre requirements, and effectiveness, while also requiring the use of advanced vocabulary and hard-to-apply grammar rules (Dobrin, 1989). Comprehending and explaining complicated information and terms necessary for tasks is also difficult for non-experts. Students' unmotivated acts or ignorance of technical tasks exemplify such hardships. As a result, the technical tasks need extra care for both instructors and students. For these reasons, intending to facilitate technical writing, this inquiry is relevant, interconnected, and significant for the investment of time and effort in research. In parallel to this, the researcher was told by his own students that they struggled to complete technical writing assignments, such as writing a recipe to a satisfactory standard. Therefore, the major drive to conduct this study is to develop ways to facilitate the planning and writing process of technical writing and ease complaints regarding the difficulty of technical writing tasks while simultaneously contributing to the related literature.

The issue to be addressed in this chapter lies in whether challenges in technical writing are related to the efficiency of error analysis. If the answer is affirmative, how and why are all of these components, *error analysis*, *process-oriented writing*, and *social constructivism*, related? As a result, because the study identifies and explains errors in technical writing works, it combines the best characteristics of relational and explanatory case studies. The concept of interest in the study is L2 learners' linguistic development, with the intention of bridging the gap in the existing literature, as no previous study has examined the confluence of these three domains. Accordingly, with this study objective in mind, the research problems were identified as follows:

RQ 1. How are Error Analysis, Process-oriented writing, and Social Constructivism attributed to the quality of technical writing?

RQ 2. Why do Error Analysis, Process-oriented writing, and Social Constructivism all complement one another in terms of technical writing achievement?

Sampling

Having formulated research questions, the entity constituting this case study was 13 English language learners enrolled in a Pre-Sessional English summer program in Turkey; for sampling selection, the researcher benefited from the purposeful sampling of those who are familiar, easily accessible, and willing to provide informed consent (Emmel, 2013). In regard to the exact number of participants required, there is no predetermined number for carrying out case studies; however, numerous resources, from articles to books and guidelines, suggest a number between 5 and 50 participants being adequate (Morse, 2000). Depending on the quality, scope, and nature of the research topic, these 13 volunteers (refers to the whole group) fit into the present study. This certain number also saturated the study, as the sample explained, supported evidence, and provided repeated justification for the arguments.

This group was a B1 repeat class in summer school and was observed to become highly unmotivated when performing technical writing tasks because the complexity was a burden to them. To examine the implementation and impact of error analysis aligned with process-oriented writing and social constructivism on technical writing in an English preparatory school, the researcher established a framework for achievement in technical writing.

The curriculum at the time consisted of just one technical writing topic at this school; creating recipes. With the extension of the present study, the researcher designed a 5-week syllabus centered on technical writing and its primary characteristics and most common types. Each type was initially introduced to the class with writing samples; they then discovered the organization/layout of the sample text and also gained insight into the related grammar topics and vocabulary in a deductive way.

As all 13 students in the study volunteered to take part in the research, the researcher's selection criteria were straightforward: the participants were the researcher's own students who shared the researcher's background, experience, level of English proficiency, and social skills, all of which are positive features for research in L2 teaching. The researcher investigated the phenomenon of interest in such a natural context, in the classroom, therefore maximizing the ecological validity of the investigation. However, for ethical issues, the real names of participants were not used to protect their anonymity. Here is the action plan for the study:

1. A technical writing task prior to the intervention (pre-test)

2. Inter-rater reliability between markers
3. A one-month technical writing study: all the participants took part.
4. The implementation of the same technical writing topic (post-test)
5. Interviews

Data Collection

The researcher used multiple data collection sources, which enabled him to augment evidence from different sources (Morgan, 2022). These include interviews, document analysis, and random participant observation on the spot. Document analysis (task papers) on students' writing samples is deemed crucial to demonstrate progress and, by extension, to provide explanations for research questions. Table 1 presents the timeline for the data collection process.

Table 1. Summary of database

Methods	Data Collection Period	Data
Document Analysis Task prior to intervention (Pre-test) User's handbook for air fryer	1st week	Teachers' evaluation for writing papers
Class Observation	Ongoing	Field notes
Training:		
Text Analysis Instruction Manuals (User's manual for a laptop & kettle)	1st week	Handouts for sample text Peer> Teacher> Self-evaluation
Text Analysis Recipes (Apple Pie)	2nd week	Handouts for sample text Peer> Teacher> Self-evaluation
Text Analysis Stages of a Process Diagrams (How plastic is recycled)	3rd week	Handouts for sample text Peer> Teacher> Self-evaluation
Text Analysis Company Documents (Business plan for a daily home catering company)	4th week	Handouts for sample text Peer> Teacher> Self-evaluation
Document Analysis Task following intervention (Post-test) User's handbook for air fryer	5th week	Teachers' evaluation for writing papers
Interviews	6th week	12 interviews total, an average of 25 minutes each

Data for this study, collected over the course of 6 weeks, encompassed pre-and post-test scores for writing tasks; document analysis (writing task papers) for error types; class observations; and complete transcriptions of in-depth interviews with 12 writing students. The diversity of data sources serves as a tool to ensure data triangulation.

Let us now separate and analyze each stage of the data collection process. The very first week's work was devoted to the pre-test process. Here, the students were tasked with creating a user handbook for the air fryer that includes installation instructions, operating instructions, and safety warnings. Some prompts related to this topic were provided to the students to get a better outline, but no other assistance was offered. Then, the researcher and another external marker graded the collected task papers. Within the same week, the researcher supplied a second subject with a sample writing task to read: An instruction manual for a laptop, including how-to-work it under necessary safety information.

Following intensive analysis of the material, the researcher helped them deduce the organization and style themselves and rediscover how certain grammar and vocabulary items were used in technical writing. After this training session, he organized a group of three and created an instruction manual for a kettle. In the meantime, he encouraged them to brainstorm and expand their ideas and learn from one another, which constituted the first phase of process writing: pre-writing. By working in a group, they were able to see similarities and differences between their experiences and proficiency in English and provide implicit and/or explicit teaching in carrying out the task. Having helped them outline the collected information, the researcher walked them through the phase of creating a draft.

In revision, the third stage of process-oriented writing, the students were initially instructed to go through their own work as a group and make corrections where necessary. They were then asked to do the same for another group by detecting mistakes and providing positive feedback as well. The groups exchanged ideas, constructed their own meanings for this challenging assignment, and edited their work. The final copy was handed to the researcher for a final expert review. Each group received detailed feedback for their work in a session outside the class. Yet, the following lesson started with a focus on fixing the most prevalent mistakes and having self-reflective discussions on the task they completed.

The same process was applied to the second (Recipes), third (Stages of a Process/ Diagram), and fourth (Company Documents) weeks, with the exception that the group study was limited to pair work in the final two weeks so that they could develop into self-sufficient, autonomous writers over time. Through socially engaging group work and constructive feedback, error analysis was developed within the frameworks of both process-oriented writing (Seow, 2002) and social constructivism (McKinley, 2015) over four hours of instruction per week.

As for the interview, a semi-structured interview with 12 (as one student could not join) students was conducted face-to-face and was recorded on the phone after getting initial consent from each participant. The recordings were transcribed and entered into word documents.

Data Analysis

That the researcher was also the writing instructor meant that he provided intimate familiarity with the study context. Such a connection, however, may work for or against the researcher. To strike a balance between objectivity and subjectivity, the researcher sought assistance from an external marker for the task paper grading process in order to maintain objectivity. To ensure rater-reliability between the researcher and the second-rater, the first-class task was subjected to Cronbach's alpha, with the result (0.802 $p<0.001$) indicating an outstanding agreement (Holcomb & Cox, 2017) between both, allowing them to go further. The fourth-week post-test followed the same pattern. The produced papers for the technical writing tasks from weeks I–III were not assessed through grading, as their primary purpose was to serve as a tool for training rather than a means of assessment.

Class observations also turned out to be positive over time, rather than unpleasant, as evidenced by students' attitudes and observed behaviors, relevant linguistic data use, comments, and class engagement. These observations were made without the use of any specialized equipment; additionally, gestures and nonverbal interaction forms were also taken into account.

As for the interviews, after the irrelevant parts and redundancies were taken out, the transcriptions were subjected to a qualitative content analysis by the researcher and a second expert to produce an organized, brief, and meaningful form out of the verbal texts (Creswell, 2007). After each side shared their individual sets of codes, themes, and categories, they were able to settle on a consensus without conducting further investigation.

RESULTS

After each pair or group completed their final copy for the task, they handed it, at first, to another group for inspection, and then took it back for revision, as noted earlier as a requirement of process writing. At the end of the class, the researcher gathered all the written works to detect errors. As also shown in Figure 1, the following paragraphs detail how the errors in these tasks were analyzed according to the phases outlined by Corder.

Figure 1. Stages of error analysis

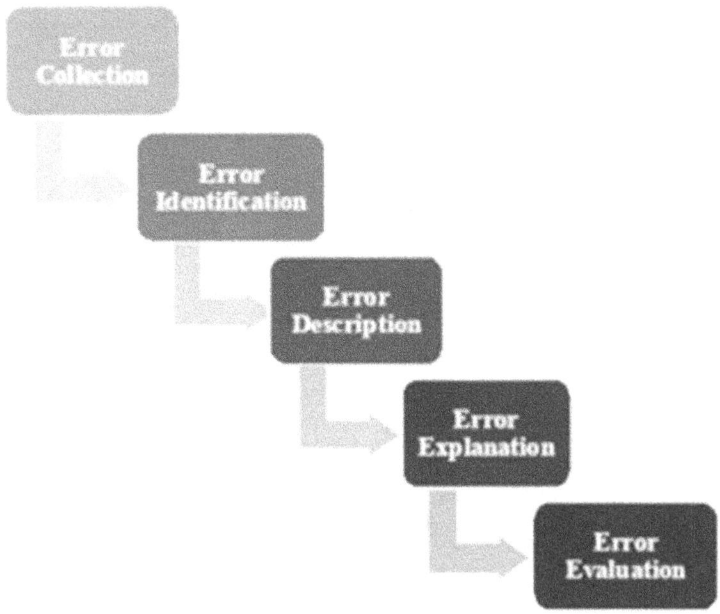

Gathering the students' works formed the initial phase of data collection and serves as a base. The researcher then identified what the errors were, as he was going through each paper, and thus created a pool of relevant errors. Next, he classified the errors into several distinct categories. The error-categories revealed in the present study are indicated in Table 2.

Table 2. Error description matrix

	Micro-textual Errors		Macro-structural Errors		
	Vocabulary	Grammar	Content	Discourse	Style
Word Level	*Improper terminology *Wrong word use *False collocations	*Incorrect use of prepositions *Determiners			
Sentence Level	*Incorrect form of the word *Inconsistent use of a word	*Passive voice *Lack of language variety *Syntactical incorrect word order *Omission of 'to be'			
Text Level		*Lack of cohesive elements	*Poor content *Redundancy	*Lack of coherence	*Informal tone *Unorganized style

Over the course of the intervention, the errors labeled with regard to the dimensions of linguistic systems fall under two categories, as displayed on the horizontal axis: micro-textual errors, referring to the flaws in vocabulary and grammar; and macro-structural errors, the ones beyond text level, which are discourse, content, and style (Hamburger, 1981). On the other hand, on the vertical axis, errors are classified into three levels in terms of the developmental layers of language: word-, sentence-, and text-levels. The most common errors are listed in descending order, ranking from highest to lowest; however, because the errors concerning macrostructure, such as informal tone, are one whole thing, they cannot be ranked individually and shown directly in the frequency table, rendering such a display irrelevant.

Initially, it is imperative to note that the greatest challenge for the participants under study was to have limited vocabulary to express their ideas in their work. As their dictionary use cannot be viewed as an error, the analysis presents an error explanation by classifying the potential source of the error into three main error categories: interlingual, which is attributed to the native language; intralingual errors, which occur because of the language being learned. Another source is developmental errors, occurring when learners build up their competence but still have limited experience and exposure to the target language (Keshavarz, 1994). Table 3 indicates the error types, the examples, and their sources in a clear way.

Table 3. Errors and their sources

	Examples	Source
Vocabulary		
Improper Terminology	make the dough bigger, *instead of* Let it sit rise for 30 minutes when it warms up, *instead of* once it heats take it to service, *instead of* contact service pour the food, *instead of* the mixture boil it, *instead of* bring it to boil	Developmental error
Wrong word use	open the air fryer, *instead of* turn on time finishes, *instead of* is over	Interlanguage
False collocation	kitchen devices, *instead of* kitchen appliances win trust, *instead of* earn trust direct a business, *instead of* run business	Interlanguage Intralanguage
Incorrect form of the word	electric shock, *instead of* electrical shock	Interlanguage
Inconsistent use of a word according to context	turn the wanted language, *instead of* the desired language	Interlanguage
Grammar Errors		
Incorrect use of prepositions	press 15 seconds, *instead of* press for 15 seconds passes through the heater, *instead of* into the heater in 350 degrees, *instead of* at 350 degrees	Interlanguage Intralanguage
Determiners	go to the setting, *instead of* go to setting	Interlanguage
Passive voice	The supply cord is damage *instead of* damaged is power by the system, *instead is* powered by	Developmental error
Lack of language variety	Overuse of 'must' or 'should' *instead of* Ensure, make sure that During before after using, *instead of* also during use	Developmental error
Syntactically incorrect word order	Enjoy what have you just created, *instead of* enjoy what you have created to not lose, *instead of* not to lose	Intralanguage

Continued on following page

Table 3. Continued

Omission of 'to be'	The company committed to ... *instead of* the company is committed to	Interlanguage
Lack of cohesive elements	Constant use of next, then, *instead of* the next stage, following this Customers have the right ... customers also ... *instead of* using they, those, the ones, etc Pour some sugar to the mixture. The mixture is ... *instead of* this mixture is ...	Developmental error
Content Errors		
Poor content	Overuse of short sentences Mix the dough, then put it into the oven Put the pie in the fridge. And turn on the oven, and wait till it is ready hot. *instead of* Turn on the oven, chill the pie while the oven heats up	Developmental error
Redundancy	add ... flour into the bowl, and add sugar ... *instead of* put flour along with sugar ...	Developmental error
Discursive Errors		
Lack of coherence to connect ideas	... healthy nutrition and catering have been enormous during the pandemic. The takeaway and catering will expand in the near future, *instead of* similarly, the takeaway and catering	Developmental error
Style Errors		
Informal tone	Our company will be the number one ... , *instead of* ... aims to rank among the top enterprises ... and this is our guidelines to run ... *instead of* with this set-mind, we will run ...	Lack of attention Developmental error
Unorganized text style	If you are caught, ... *instead of* if one is caught	Lack of attention Developmental error

The study aimed to analyze the errors related to grammar, vocabulary, content, discourse, and style, leaving spelling out as the research's purview was already so intensive. However, we will now examine major error types for each category. For vocabulary, the greatest challenge was the students' vocabulary incompetence for certain technical writing topics. Although the researcher provided one or two writing samples to them, such an effort was still inadequate, as technical writing is highly dependent on the use of specialized jargon. This kind of error is attributed to the developmental language; therefore, the source of the error is described as intralanguage and developmental errors. The students' poor vocabulary knowledge, along with English's extensive vocabulary, could explain this type of deficiency.

As for grammar errors, the use of prepositions accounts for a substantial share of all grammatical errors. We can attribute these slips to Turkish's shallow mentality on the topic of prepositions, which results in more inconsistencies in the students' sentences in English. The structural feature of Turkish that allows one to say 'press the button 15 seconds' causes negative transfer into English. The fact that prepositions pose a great challenge for some L2 learners can be explained by the principle that the degree of difference between two languages is proportional to the degree of difficulty (Lado, 1957).

As one subdomain under macro-structural errors, content posed a serious threat to L2 learners. As the study group was just pre-sessional English program students, they did not have any of the required field expertise, specific terminology, or relevant life experiences. Because of their blank minds on the assigned technical writing topics, they had great difficulty coming up with ideas, designing content, and determining the flow, leaving us with near-satisfactory writing content.

Concerning discourse, the produced texts lacked coherent and cohesive elements that would connect ideas for readers. For sentences and thus ideas to flow smoothly through cohesive elements and linking words, the students initially performed at an average level but, in time, advanced this skill to a better level. The prominent problem related to discourse is that L2 learners left some overall gaps in mental flow, meaning that some constituents were missing in forming a whole. To illustrate this, in the tasks of a recipe, such parts as the stage of preheating the oven to what degree or how to weave dough strips on both sides of the pie are left with no instructions. As any technical writing piece should be specific and consist of all relevant details, the items should be written in a way necessary to meet the demands of the task. For example, in the task of writing a business proposal for a daily home catering company, many students made the mistake of addressing one certain target group, instead of their intended audience. Beginning with such a start meant skipping some other details, such as competitive advantage, planning, cost, and the rest.

Less drastic measures were needed to fix problems related to style than the other components of macro-structural errors. The most significant barrier, in this case, was the informal nature of the presented work. For example, again in the same task of business proposal, a group wrote;

'Our company will be the number one daily home catering company.'

This informality may sound arrogantly self-esteemed to the reader and violate the conventions of the genre of technical writing.

Having detailed the first four stages of error analysis, in the phase of error evaluation, the researcher allocated some time to explicit feedback related to errors in order to remediate them further. The change in class average scores from the pre- to post-test was used to evaluate the effectiveness of the technical writing intervention. The task of a user handbook for air fryer was assigned to thirteen students in the first and fifth weeks of the program to determine whether the intervention assisted L2 learners in acquiring the foundation of technical writing.

As for the feedback, peer feedback was observed to be more focused on grammar and vocabulary, which were easier to detect and correct, while the teacher was still responsible for addressing the remaining micro-textual errors in these parts, and also for helping L2 students break the codes of macro-structural errors related to the content, discourse, and style. The errors at the micro-textual and macro-structural levels were separately evaluated by the researcher and the second inter-rater. Regarding the difference between the averages for micro-textual components, grammar, and vocabulary, the students made progress by 38%, a jump from 42 to 58 on a scale of a possible 100. However, macro-structural elements could not be scaled up in the same way as micro-textual ones could. Class progress increased from 38 in the beginning to 47 out of 100 in the end, for a 24% better performance. The bar chart in Figure 2 presents the average class grades before and after the test.

Figure 2. Class average grades for pre and post tests

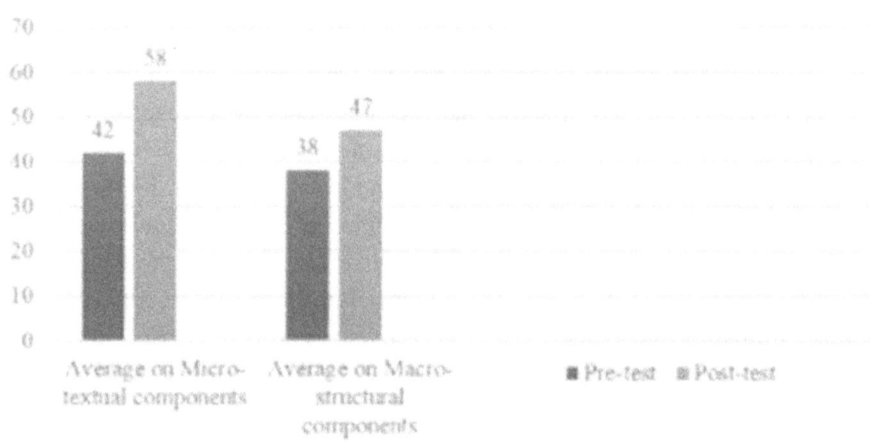

The initial grades are relatively low since they reflect their below-average technical writing competence and performance. The most likely explanation is that they were not given enough of a challenge in technical writing or did not have any intensive training before. Meanwhile, the difference in progress between micro-textual and macro-structural levels lies in the rationale that grammar and vocabulary are tangible skills that can explicitly be seen, corrected, and easily acquired. However, designing content, creating a coherent and cohesive flow, and composing within the parameters established by the conventions of the genre are all very intangible and elusive skills to acquire. And creating a product in each subgenre of technical writing (recipes, business plans, manuals, etc.) is viewed as evidence of mastery of skills in that discipline, along with the right terminology and writing guidelines. For these reasons, the progress in the macrostructure domains may have indicated a lower level of performance than in the micro-textual ones.

Lastly, a semi-structured face-to-face interview was conducted to gather information about participants' perceptions, attitudes, ideas, suggestions, and behaviors in relation to technical writing and the intervention. This was accomplished through an interview protocol consisting of ten questions on the intervention and their general perceptions of technical writing, process-oriented writing, and social constructivist approach to writing and the combination of these three. The one-on-one interview adhered to the protocol (Castillo-Montoya, 2016) as follows:

Interview Protocol Stages

1. *Background/introductory questions*: To warm up in a general and non-threatening format
2. *Transition questions*: To deliver a transfer between introductory and key questions
3. *Key questions*: To reveal core questions related to research objectives
4. *Closing questions*: To give participants the chance to add some other points

The transcriptions obtained from the interviews were analyzed through content analysis separately by the researcher and another expert colleague. As indicated in Table 4, the codes and themes were labeled and categorized into the following groups:

Table 4. Perceived underlying factors in the technical writing intervention

		Themes	Participants (n=12)												Total
			Male					Female							
			P1	P3	P7	P9	P10	P2	P4	P5	P6	P8	P11	P12	
Positive	Facilitators	Scaffolding		•	•	•	•		•	•	•	•	•	•	83%
		Relatedness	•		•		•	•	•		•			•	55%
	Outcomes	Increased confidence		•	•			•		•	•	•	•	•	67%
		Gradual development in writing skills	•	•		•	•	•		•	•		•		67%
Negative	Hindrances	Task Difficulty	•	•	•	•	•	•	•	•	•	•	•	•	100%
		Limited Technical Vocabulary		•			•		•	•		•		•	50%

Factors both beneficial and detrimental to technical writing and intervention were discovered as two dominant themes. The researcher's provision of writing samples in class to aid participants' understanding of the tasks was an asset to the process because the samples served as a blueprint for the work and assisted participants in recreating an identical one. Ten students made references to receiving professional assistance, which was labeled as 'scaffolding' since it provided support to improve learning and mastery as students were acquiring new skills. This demonstrates the significance of sample writing and its analytical de-structural progress before creating a new one. Likewise, participating in a cooperative effort to accomplish a goal fosters the growth of caring relationships and a stronger sense of belonging. In light of this, the content analysis found that a sense of relatedness was the second most prevalent theme, showing a sign of L2 learners having yielded better performance in the writing class. They were able to clearly observe the relief and content that were brought to them through the intervention at the end of the sessions, and they reported feeling more confident about themselves and their writing skills as a result of their successes in these challenging technical writing tasks. A similar number of participants cited increased self-confidence as a key component of the process, as did an increase in technical writing competence and performance. This once again underscores the significance of psychological factors and emotional support in second language learning (Sartika, 2022). Regarding the factors hindering the process, it is observed that task difficulty, which all participants mentioned and even complained about, and their limited vocabulary while doing the tasks, are two variables that hampered the process and perceptions of technical writing. From the student's perspective, these results indicate that vocabulary has a crucial role in writing compared to grammar; that's why more sessions should be allocated to technical vocabulary acquisition.

SOLUTIONS AND RECOMMENDATIONS

With regard to the research questions, the general outcome of the intervention of a process-oriented social constructivist approach to technical writing was beneficial for both planning and writing the tasks, as well as providing emotional support to handle such demanding assignments. L2 students made significant progress in micro-textual error correction, grammar, and vocabulary. The students' 38 percent improvement in technical writing tasks may not seem like a great leap, but the fact that the students were inexperienced in writing due to their lack of background knowledge justifies such progress as noteworthy. The advancement in macro-structural components was a bit sluggish as the competence of technical content, discourse, and style is extremely enormous and hard to acquire. Yet, a quarter's overall progress (24%) should still be assessed as a great triumph, instead of L2 students giving the teacher the cold shoulder, even when they were just presenting the task.

Due to the fact that English includes anywhere between 60 and 70 prepositions, practically all of which have distinct and diverse meanings, the error analysis for grammar found that the topic of prepositions comprised the trickiest component of the tasks, forming the majority of errors. Therefore, prior to technical writing work, some teaching sessions should be allocated to enable them to acquire the most commonly used ones and more complex phrases of prepositions.

The same applies to determiners (a, an, the), as some languages have different word orders or possess no 'the' at all, and they are ideally omitted in technical writing. The sources of these errors stem from native language interference and the complexities of English as the target language. Additionally, the study suggests that the students' assumption that there are no variations between the two languages may be at the root of poor linguistic performance on these tasks.

The major handicap related to poor content was due to the task's difficulty and their unfamiliarity with technical writing assignments. It was observed that limited vocabulary was a great negative factor in creating dedicated content in terms of technical outputs. To express this hardship, one interviewee stated,

I never thought English has this much vocabulary, especially in technical writing. I thought I knew lots of words, but seemingly, it is never enough, especially for these kinds of writing tasks.

This statement takes us to the conclusion that vocabulary is one major component in technical writing, as it is a great messenger that leaves nothing out for target readers to comprehend the writing.

The content analysis found that technical writing is perceived as a 'very difficult type of writing' as it requires advanced vocabulary and grammar structures and includes distinct structural, discursive, and stylistic elements. Therefore, helping hands should provide both instructional and emotional support for L2 technical writing students; otherwise, it would only lead to an inefficient individual struggle in vain. The sample text analysis, in which the teacher de-structured the text into rational sections, was found to be effective. Keeping in mind that macro-structural components function in alignment with discursive markers and required vocabulary helps L2 writers restructure or, if possible, duplicate the same constituents. This forms a scaffolding that gives students handy assistance in producing something totally new. As expressed by 58% of the interview participants, collaborative work made them feel related to each other, increasing their sense of belongingness, academic motivation, task efficiency, and even their psychological well-being (Kurt, 2021). The content analysis also revealed the significance of emotional support and the feeling that one achieves a hard-nut challenge, which in turn creates confidence and a sense of craving for further learning.

FUTURE RESEARCH DIRECTIONS

Upon the completion of the study, the researcher realized that a 4-week intervention that aimed to improve technical writing skills was not long enough to acquire the competence and practice necessary. Besides, one takeaway assignment or two was not enough to grasp the conceptual and practical aspects of one single type of technical writing; thus, at least four or five similar tasks for each type of technical writing (e.g., recipes and business proposals) would have made a significant contribution to the process. For future investigations to make a more comprehensive view and cover more ground, spelling may also be added to the scope of constructs to investigate.

CONCLUSION

The phrase technical writing embraces a vast range of texts in research, engineering, and skilled crafts. Because it calls for careful attention to various elements, technical writing is a difficult task for L2 learners in many ways. With an understanding of what technical writing is and how the convergence of social-constructivist process-oriented writing and error analysis works, L2 students can get an effective and easy-to-follow method for learning how to write efficiently in technical writing. This chapter set out to investigate whether L2 writers may benefit from error analysis based on process-oriented writing classes within a social-constructivist framework.

As for the types of errors in the present study, inadequate and incorrect terminology and prepositions in grammar at the micro-textual level account for the majority of vocabulary errors. As for the macrostructure, the most common errors are inadequate or limited content knowledge for content, a lack of coherence for discourse, and an informal tone for style. All of these errors stem from sources such as intralanguage and interlanguage and must be addressed properly. Despite the task difficulties in technical writing that L2 writers encounter, the assistance they receive from their peers and instructors, the sense that they belong together and are related to each other in writing, and the idea that their writing is gradually progressing all enhance their motivation and learning process.

Overall, error analysis aligned with process-orientation writing, and the social constructive approach contributes a lot to L2 students' learning because the convergence of these three gives a clear guideline and socially learning support during the process and the feedback that is both corrective and encouraging. Process-oriented writing assists the progress in significantly proving a blueprint with its sharp step-by-step outline while clearing 'clouded minds' for those incompetent L2 writers. As an added bonus, it aids in producing finished goods that are both of higher quality and more logically structured. Meanwhile, the social constructivist approach to writing has helped students build meaning together and deal with issues that are above their language capabilities. Furthermore, this method results in valuable, corrective, and encouraging feedback from both their peers and teacher, allowing them to ultimately master the tasks at hand.

The study, with its unique perspective, achieved to portray the compatible convergence of error analysis aligned with process-oriented writing and a social constructivist approach to writing. It also revealed how each component complements the other and considerably assists novice writers in composing a finished product as far as the tremendously challenging technical writing process is concerned. The study also suggests that the quality of written texts is attributed first to the language skills, task awareness, and related writing experience of L2 writers, as well as their familiarity with the necessary materials and tools. Figure 3 illustrates the cooperation between error analysis, process-oriented writing, and social constructivism.

Figure 3. Convergence of process-oriented writing, social constructivist approach and error analysis

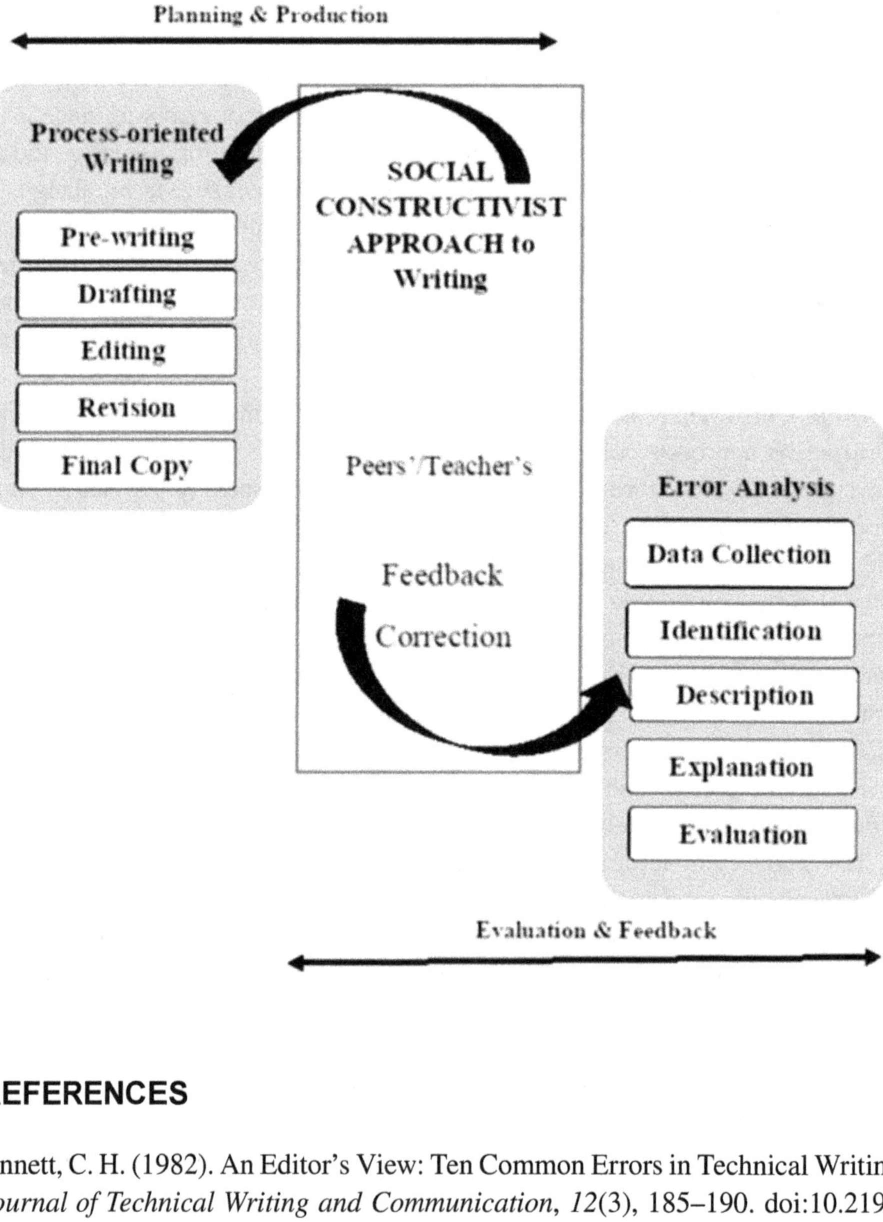

REFERENCES

Annett, C. H. (1982). An Editor's View: Ten Common Errors in Technical Writing. *Journal of Technical Writing and Communication*, *12*(3), 185–190. doi:10.2190/W558-WV3D-F1P6-HXMD

Berninger, V. W., & Winn, W. (2006). Implications of advancements in brain research and technology for writing development, writing instruction, and educational evolution. In C. MacArthur, S. Graham, & J. Fitzgerald (Eds.), *Handbook of writing research* (pp. 96–114). Guilford Press.

Britton, W. E. (1965). What Is Technical Writing? *College Composition and Communication, 16*(2), 113. doi:10.2307/354886

Brown, D. H., & Lee, H. (2001). *Teaching by principles: An interactive approach to language pedagogy*. Pearson Education ESL.

Brown, H. D. (2001). *Teaching by Principle: An Interactive Approach to Language Pedagogy*. Longman.

Castillo-Montoya, M. (2016). Preparing for Interview Research: The Interview Protocol Refinement Framework. *Qualitative Report*. Advance online publication. doi:10.46743/2160-3715/2016.2337

Corder, S. P. (1967). The Significance of Learner's Errors. *IRAL -. International Review of Applied Linguistics in Language Teaching*, *5*(1–4). Advance online publication. doi:10.1515/iral.1967.5.1-4.161

Creswell, J. W. (2002). *Research Design: Qualitative, quantitative, and mixed methods approaches* (2nd ed.). SAGE Publications.

Dobrin, D. N. (1989). *Writing and Technique*. Natl Council of Teachers of English.

Emmel, N. (2013). *Sampling and choosing cases in qualitative research: A realist approach* (1st ed.). SAGE Publications Ltd. doi:10.4135/9781473913882

Eunson, B. (2011). Communicating in the 21st Century 3E iStudy Version 1 (3rd ed.). Wiley.

Flowerdew, J. (2016). English for Specific Academic Purposes (ESAP) Writing. *Writing & Pedagogy*, *8*(1), 1–4. doi:10.1558/wap.v8i1.30077

Galbraith, D. W. (2012). *Writing about what we know: Generating ideas in writing*. SAGE Publications Ltd. doi:10.4135/9780857021069.n4

Gill, A. E., & Janjua, F. (2020). Genre pedagogy and ELLs' writing skills: A theme analysis. *English Language Teaching*, *13*(8), 141. doi:10.5539/elt.v13n8p141

Hamburger, H. (1981). *The Function of the Predicate in the Fables of Krylov: A Case Grammatical Study*. Rodopi.

Holcomb, Z. C., & Cox, K. S. (2017). *Interpreting basic statistics: A workbook based on excerpts from journal articles*. Taylor & Francis. doi:10.4324/9781315225647

Hyland, K. (2003). *Second Language Writing* (1st ed.). Cambridge University Press. doi:10.1017/CBO9780511667251

Hyland, K., & Hamp-Lyons, L. (2002). EAP: Issues and directions. *Journal of English for Academic Purposes*, *1*(1), 1–12. doi:10.1016/S1475-1585(02)00002-4

James, C. (1998). *Errors in Language Learning and Use: Exploring Error Analysis (Applied Linguistics and Language Study)* (1st ed.). Routledge.

Karney, B. W., & Filion, Y. (2000). Technical writing: Purpose and principles. *Journal of Hydraulic Engineering*, *126*(4), 230–231. doi:10.1061/(ASCE)0733-9429(2000)126:4(230)

Keshavarz, M. H. (1994). *Contrastive analysis and error analysis*. Rahnama Publications.

Kroll, B. (1990). *Second language writing: research insights for the classroom*. Cambridge University Press. doi:10.1017/CBO9781139524551

Kurt, A. (2021). *Self-Determination and its relationship to conscientiousness, career motivation, gender, and the year of study in pre-service English language teachers* (Publication No: 677265) [Doctoral Dissertation, Onsekiz Mart University]. YÖK Ulusal Tez Merkezi. https://tez.yok.gov.tr/UlusalTezMerkezi

Lado, R. (1957). *Linguistics Across Cultures: Applied linguistics for language teachers*. University of Michigan Press.

Larissa, B., & Valeria, C. (2019). Technical Writer in the Framework of Modern Natural Language Processing Tasks. *Journal of Siberian Federal University. Humanities & Social Sciences*, *12*(1), 20–31. doi:10.17516/1997-1370-0377

Leong, L. M., & Ahmadi, S. M. (2017). An analysis of factors influencing learners' English speaking skill. *International Journal of Research in English Education*, *2*(1), 34–41. doi:10.18869/acadpub.ijree.2.1.34

McKinley, J. (2015). Critical Argument and Writer Identity: Social Constructivism as a Theoretical Framework for EFL Academic Writing. *Critical Inquiry in Language Studies*, *12*(3), 184–207. doi:10.1080/15427587.2015.1060558

Morgan, H. (2022). Conducting a Qualitative Document Analysis. *Qualitative Report*, *27*(1), 64–77. doi:10.46743/2160-3715/2022.5044

Morse, J. M. (2000). Determining sample size. *Qualitative Health Research, 10*(1), 3–5. doi:10.1177/104973200129118183

Nunan, D. (1991). *Language teaching methodology: A textbook for teachers.* Longman.

Nurfaidah, S. (2018). Vygotsky's Legacy on Teaching and Learning Writing as Social Process. *Langkawi: Journal of the Association for Arabic and English.* doi:10.31332/lkw.v4i2.1038

Peter, J., & Singaravelu, G. (2021). Problems in writing in English among high school learners. *AEGAEUM Journal, 8*(9). 1502-1515. https://www.researchgate.net/publication/351048985_Problems_in_Writing_in_English_among_High_School_Learners

Powell, C. K., & Kalina, J. C. (2009). Cognitive and social constructivism: Developing tools for an effective classroom. *Education, 130*, 241–250. doi:10.4236/ojapps.2022.121011

Priya, M. R., & Divya, K. V. (2016). Writing skills for technical writers. *International Journal of Engineering Technology, Management and Applied Sciences, 4*(2). https://www.researchgate.net/publication/304150238_Writing_Skills_For_Technical_Writers

Rana, L. B. (2018). *L1 Use in L2 Writing: A Case of Multilingual Undergraduates in a Basic Writing Course* [Master's thesis, Minnesota State University, Mankato]. Cornerstone: A Collection of Scholarly and Creative Works for Minnesota State University, Mankato. https://cornerstone.lib.mnsu.edu/etds/786/

Rao, V. C. S. (2020). Academic language proficiency: English for Academic Purposes (EAP). *Journal for Research Scholars and Professionals of English Language Teaching, 21*(4). https://files.eric.ed.gov/fulltext/ED611477.pdf

Reid, J. M. (2001). The Cambridge guide to teaching English to speakers of other languages. In R. Cater & D. Nunan (Eds.), *Writing* (pp. 23–33). Cambridge University Press.

Richards, J. C., & Renandya, W. A. (2002). *Methodology in language teaching: An anthology of current practice.* Cambridge University Press., doi:10.1017/CBO9780511667190

Sağlamel, H., & Aydoğdu, Z. M. (2021). The Academic writing needs of students: A case study on stakeholder perspectives. Acuity. *Journal of English Language Pedagogy, Literature and Culture, 7*(1), 37–50. doi:10.35974/acuity.v7i2.2541

Saito, K., & Hanzawa, K. (2018). The role of input in second language oral ability development in foreign language classrooms: A longitudinal study. *Language Teaching Research*, *22*(4), 398–417. doi:10.1177/1362168816679030

Sartika, D. (2022). A Correlation between Students' Self-Esteem and Speaking Skill. *JETLe*, *3*(2), 1–9. doi:10.18860/jetle.v3i2.15610

Sarwat, S., Ullah, N., Anjum, H. M. S., & Bhuttah, T. M. (2021). Problems and factors affecting students English writing skills at elementary level. *Elementary Education Online*, *20*(5), 3079–3086. doi:10.17051/ilkonline.2021.05.332

Schreiber, L. M., & Valle, B. E. (2013, May 29). Social Constructivist Teaching Strategies in the Small Group Classroom. *Small Group Research*, *44*(4), 395–411. doi:10.1177/1046496413488422

Selinker, L. (1972). Interlanguage. *International Review of Applied Linguistics in Language Teaching*, *10*(1–4). Advance online publication. doi:10.1515/iral.1972.10.1-4.209

Selinker, L. (1992). *Rediscovering Interlanguage*. Longman.

Seow, A. (2002). The Writing Process and Process Writing. In J. C. Richards & W. A. Renandya (Eds.), *Methodology in language teaching* (pp. 315–327). Cambridge University Press. doi:10.1017/CBO9780511667190.044

Touchie, H. Y. (1986). Second language learning errors their types, causes and treatment. *JALT Journal*, *8*(1), 75–80.

Vygotsky, L. S. (1978). *Mind in society: The development of higher psychological processes*. Harvard University Press.

Vygotsky, L. S. (1986). *Thought and language*. MIT Press.

Wilson, K., & Devereux, L. (2014). Scaffolding theory: High challenge, high support in Academic Language and Learning (ALL) contexts. *Journal of Academic Language and Learning*, *8*(3), 91–100. https://journal.aall.org.au/index.php/jall/article/download/353/210

Zhou, Y., & Wei, M. C. (2018). Strategies in technology-enhanced language learning. *Studies in Second Language Learning and Teaching*, *8*(2), 471–495. doi:10.14746sllt.2018.8.2.13

KEY TERMS AND DEFINITIONS

Error Analysis: A process-oriented method that helps L2 learners recognize, evaluate, and classify the deviations that occur between their language use and the norms of the target language.

Process-Oriented Technical Writing: A technique that focuses on the process rather than the end product, allows L2 learners to understand how to work through and discover strategies related to how authors typically create.

Social Constructivism: An approach that learning is an active process by which learners actively create meaning out of knowledge, unlike a passive orientation in which they are merely handed information.

Chapter 3
Error Analysis of Teacher Technical Writing:
The Corpus of EFL Lesson Plans

Banu Çiçek Başaran Uysal
https://orcid.org/0000-0002-4978-0891
Eskisehir Osmangazi University, Turkey

EXECUTIVE SUMMARY

This chapter delves into an investigation of pre-service English as a foreign language teachers' technical writing products, by employing the Surface Strategy Taxonomy proposed by Dulay et al. (1982). By shedding light on the written errors of future English language teachers, this study contributes to our understanding of their language learning needs and presents implications for improving teacher education practices. The chapter examines a corpus of lesson plans written by senior students (n=140) enrolled at a public university, aiming to become English language teachers. Lesson planning can be considered a form of technical writing for language teachers since it requires adherence to specific techniques and the use of appropriate language structures to effectively convey in-class procedures. The analysis reveals that the prospective teachers produced erroneous utterances falling into four distinct categories: omission (45%), misinformation (30%), addition (17%), and misordering (7%). These findings serve as a foundation for several suggestions to enhance teacher education.

DOI: 10.4018/978-1-6684-6222-5.ch003

INTRODUCTION

Language users produce incorrect samples of language whether it is L1 or L2, written or spoken communication (Corder, 1967; Erdoğan, 2005). Based on the feedback they receive; the language users can improve their knowledge of target language grammar rules. Language learner errors are defined as deviations from the target language norms (Brown, 2014; Corder, 1973; Dulay, Burt & Krashen, 1982; Ellis, 2008). When talking about language learning context, it is imperative to distinguish errors from mistakes: While the former stems from a lack of knowledge of the learners, the latter refers to performance deviations due to a lack of attention, tiredness, or carelessness (Richards & Schmidt, 2010). Ellis (1997) suggests to practically identify whether the ill-formed utterance is an error or a mistake, we need to look at the performance of the learner: if they sometimes employ the correct form and sometimes the deviant form then it can be categorized as a mistake; on the other hand, if they use the incorrect form all the time, then it is an error. Language learners' errors are observed to be systematic and thus they indicate the learners' advancement in the target language (Montrul, 2011).

Identifying and analyzing language learners' errors provides valuable insights into their interlanguage, which is the language system that learners create as they develop their proficiency in a target language (Atmaca, 2016; Corder, 1975). This process is essential for language teaching, planning, and design (Richards & Schmidt, 2010). Error analysis (EA), as defined by Corder (1967), is a technique used to evaluate the speech or writing performance of EFL learners. Brown (2014) clarified EA as a reflective approach to the status of the learners' understanding of a remedial method in constructing structures for language learners, while Al-Sobhi (2019) debated the importance of EA. The literature suggests that EA can provide guidelines for instructors and syllabus designers in designing remedial language courses (Al-Sobhi, 2019; Corder, 1967; Gass, Behney, & Plonsky, 2020). For example, by following one of the methods proposed by Dulay et al. (1982), teachers could analyze students' frequent errors, understand their current level of language learning, and then use these errors as a learning tool to create pedagogical strategies to improve their language ability.

It is crucial to examine the technical writing errors of future English as a Foreign Language (EFL) teachers since these errors provide insight into the participants' target language proficiency (Almusharraf & Alotaibi, 2022; Dulay et al., 1982; Uzdu Yıldız & Çetin, 2020). Analyzing lesson plans is also important for understanding the target language needs of prospective EFL teachers. In other words, understanding the target language competency of future EFL teachers is essential for making inferences about their subject matter knowledge (Berry, Depaepe, & van Driel, 2016) or content knowledge (Shulman, 1987). Given that pre-service teachers are the future English instructors who will teach the target language to future generations, identifying the gaps in their interlanguage is vital for teacher development and language planning. Therefore, the analysis of learners' errors is critical for the successful teaching and learning of a target language.

Creating lesson plans can be considered technical writing for teachers since they need to use specific terminology and certain set of grammatical structures to refer to in-class procedures and learning outcomes. Lesson plans are of great significance for teachers since they provide a blueprint for implementing teachers' beliefs, cognition, teaching philosophy, and knowledge base into the lesson (Farrell, 2002). Lesson planning helps us identify the thought procedures of teachers as professionals and how they view language teaching. When writing lesson plans, the teachers need to concentrate on (1) the learning outcomes of the lesson, (2) prerequisite knowledge of language learners, (3) anticipated problems and their solutions, (4) classroom interaction patterns, (5) stages of the lesson, (6) classroom management, (7) in-class procedures, and (8) timing of activities. Lesson planning is not just about writing a series of steps that need to be followed in one class, but it also entails analyzing the background and proficiency level of the students, finding and choosing the appropriate language teaching techniques and activities to teach the topic, and synthesizing their professional knowledge base to build a bridge between theories and practice of language teaching (Purgason, 2014; Velliaris, 2016). It is also important to note that lesson plans not only offer teachers a road map for the upcoming class but also enable them to obtain structured feedback about the lesson at hand (Purgason, 2014). After completing the class, the teachers can revisit their plans and check if the intended activities were effective, if not they can make modifications to the plan. Owing to the importance of lesson planning for teachers, during their undergraduate studies, prospective teachers are trained for writing detailed and well-structured lesson plans. Because lesson planning is such an essential component of effective teaching, prospective teachers are trained during their undergraduate studies to write detailed and well-structured lesson plans. By polishing their technical writing skills and focusing on the necessary elements of lesson design, teachers can create effective learning experiences for their students and help ensure their success.

Despite the importance of lesson planning, there is a gap in corpus-based analysis of errors in pre-service EFL teachers' technical writing. The literature offers several pieces of research on analyzing English language learners' errors in written (Chan, 2010; Connolly, 2020; Fumero & Wood, 2021; Lee, 2020; Nikolenko et al., 2021; Sun, 2017), and spoken forms (Najlah & Fatimah, 2020; Rehman et al., 2020; Ting, et al. 2010). There are also a few studies about pre-service English language teachers and error analysis (Arıkan & Yılmaz, 2019; Aufa & Syarif, 2020; Hunt-Gómez & Navarro-Pablo, 2020; Widya Hapsari, 2021). Numerous inquiries also focus on corpus-based error analysis of English language learners (Almusharraf & Alotaibi, 2022; Castillo Rodríguez & López Pérez, 2019; Crosthwaite, 2018; Lelakova & Toman, 2023). However, the gap in corpus-based analysis of errors in pre-service EFL teachers' technical writing remains. For this reason, this study aims to provide insights into future EFL teachers' language errors by scrutinizing their professional writing practices. The EA presented in the current study provides one of the first investigations into a corpus of English teaching lesson plans written by pre-service EFL teachers by following the Surface Strategy Taxonomy (Dulay et al., 1982).

LITERATURE REVIEW

The previous studies offer several examples for analyzing EFL learners' errors in written and spoken language. For instance, by examining the written English language errors of 32 elementary students, Atmaca (2016) concluded that the participants in the study made errors resulting from interlingual (transferring from L1 to L2) and intralingual factors (overgeneralization of L2 rules). Most of the errors were related with preposition, verb, and article use. Another study by Chan (2010) also focused on written errors of learners by investigating free-writing tasks of 387 Chinese participants. The aim of the research was to propose a taxonomy of written errors. Chan (2010) indicated that L1 influence and poor understanding of L2 structures were the two major factors that affect the L2 development of the participants. For analyzing the spoken language of the learners, Ting et al. (2010) focused on grammatical errors of 42 university students. By following the Surface Structure Taxonomy (Burt & Krashen, 1982), they examined a corpus of oral interactions recorded throughout a 14-week semester. The results of the study revealed that the majority of the errors belonged to preposition, articles, and pluralization of nouns. Furthermore, the study showed that omission and misinformation errors were the two most frequently observed categories. The research conducted by Najla and Fatimah (2020) focused on grammatical errors in oral communication by employing telephone interviews with 30 university students. Similar results were also reported by Najla and Fatimah (2020): most of the errors were related with articles, subject-verb agreement, and simple past tense and the errors occurred due to incomplete knowledge of target language structures and overgeneralization (intralingual), and native language influence (interlingual).

There are also studies focusing on prospective English language teachers and errors analysis. Hunt-Gómez and Navarro-Pablo (2020) examined a 238-minute corpus created by improvised production of the participants in the study. They identified pronunciation and intonation errors of pre-service teachers. The study suggested that by employing phonetic transcription, the participants of the study can be trained to correct their errors in the target language. Another study by Widya Hapsari (2021) investigated Indonesian pre-service teachers of English and focused on the written work of the participants. The results were similar to the previously mentioned studies, most of the written errors of pre-service teachers were related with article use and subject-verb agreement. Another research was conducted with future EFL teachers, by Arıkan and Yılmaz (2020), focused on spoken errors and concluded that L1 interference and inadequate L2 knowledge were the primary reasons of the errors.

To review, the studies mentioned above provide valuable insights into the common errors that EFL learners make. Furthermore, the studies identified similar findings for the reasons of errors: L1 influence and inadequate understanding of L2 structures. This information can be used by teachers to provide effective instruction and help learners improve their English skills. Teachers can use the information from these studies to identify the most common errors that their students make and develop strategies to help them correct these errors. For instance, providing explicit instruction target language structures, presenting opportunities for learners to practice the language by writing and speaking, giving corrective feedback can be some of the strategies that can be employed to improve the English proficiency of the learners. Additionally, teachers can use the information from these studies to create a more supportive learning environment for their students, where they feel comfortable making mistakes and learning from them.

In conclusion, the present study aims to contribute to the body of literature by focusing on the technical writing errors of prospective EFL teachers. The study followed the Surface Strategy Taxonomy (Dulay et al., 1982) to categorize the errors and examined a corpus of 1400 lesson plans written by pre-service EFL teachers.

METHODOLOGY

Study Context

This study takes place at a university in an EFL setting, where the participants are training to become English language teachers. The context of the study is defined as the expanding circle (Kachru, 1985). According to the classification of English in the world by Kachru (1985), there are three levels: (i) the inner circle, where English language is spoken as the native language, (ii) the outer circle, in which English is the second language for the speakers since it is the official language, and (iii) the expanding circle where English is a foreign language for international communication. Considering these, the participants of the study are categorized as EFL learners themselves. One of the factors that differentiates expanding circle from inner and outer circles is limited opportunities for receiving target language input to learn the language. That is to say, the learners in this setting get exposed to target language during limited hours of English lessons offered at the schools (Peng, 2019). Another one is limited output opportunities for learners. It is suggested that producing output, or using the language to convey a message, can help interlanguage development of learners (Longcope, 2009). Since the learners are in foreign language setting, they have diminished chances of using the target language outside the classroom. However, having limited input and output opportunities in their language learning setting does not necessarily entail having lower target language competence. It is also important to note that the participants of the present study were training to become English language teachers themselves. To achieve this, they need to receive high scores from nation-wide university entrance and target language exams, follow and complete the courses offered in target language at the undergraduate level, and successfully complete the teaching practicum where they need to engage in tasks that links their theoretical and practical knowledge as teachers.

The participants of the study were senior college students, who were aged between 20-22, in an undergraduate program. The program aims to train English language teachers. The students are required to complete three years of theoretical courses before they can start teaching practicum in the fourth year. The teaching practicum consists of two courses offered in two consecutive academic terms. During the first teaching practicum, the prospective teachers observe English lessons for 12 weeks (6 hours a week) and teach ten English lessons. In other words, fourth-year students both attend the English lessons offered at public and state schools and practice their teaching skills in those lessons. For this reason, they are expected to write at least ten lesson plans to accompany them in their teaching practice. A similar procedure is also followed for the second teaching practice. That is to say, before they graduate, the pre-service teachers complete 24 weeks of observing and 20 hours of teaching. As can be inferred, the participants of the study were to write at least twenty different lesson plans for the practice teaching courses. Given that, the present study aims to identify the technical writing errors of prospective EFL teachers by exploring a corpus of English lesson plans.

Study Design

Language mistake research has long been a crucial component of second language acquisition studies. Traditional approaches to identifying common errors and comprehending their underlying causes involve EA, the methodical evaluation of errors produced by second language learners. The focus on surface methods, on the other hand, has become a more important aspect of EA as a way for second language learners to make up for linguistic shortcomings. In this study, the Surface Strategy

Taxonomy proposed by Dulay, et al. (1982) was followed for categorizing the errors in the lesson plan corpus. This taxonomy proposes four categories for EA: omission, addition, misinformation, and misordering. Omissions are errors categorized by the lack of morphemes or words that are required normally in a grammatical sentence, whereas additions are errors resulting in ill-formed sentences due to unnecessary morphemes or words. Misinformation errors occur because of using the wrong form of morphemes or structures. Misordering errors, as the name suggests, stem from using morpheme(s) in incorrect order (Dulay, et al., 1982). The Surface Strategy Taxonomy was considered to be more fruitful for researchers who are interested in uncovering the cognitive processes that underlie the learners' reconstruction of the new language. It also allows researchers to observe the fact that learners' errors are indeed products of systematic reasoning, rather than "laziness or sloppy thinking" (Dulay, et al. 1982, p. 150).

This study on pre-service EFL teachers' technical writing involved a rigorous and systematic process of data collection and analysis. The study took place over a period of five years, from 2018 to 2023, and involved collecting lesson plans written by 140 senior-year students at a university in an EFL setting. The lesson plans were then collated into a corpus, which initially consisted of 1550 files. However, after eliminating any corrupt or recurring files, 1400 lesson plans were left for analysis. The process of collecting and organizing such a large amount of data was a crucial aspect of the study as it ensured that the analysis was based on a representative sample of the participants' technical writing abilities.

To analyze the corpus, the researchers randomly selected one lesson plan from each participant, which was then carefully scrutinized for technical writing errors. This process of random sampling ensured that the analysis was unbiased and reflective of the participants' overall technical writing skills. The sample size of 10% of the corpus was also determined by the principle of representativeness, which is a widely accepted method of selecting a sample that is sufficiently large and representative of the population being studied (Creswell, 2012). This approach helped to ensure the validity and reliability of the study's findings. By following the established research principles, such as random sampling and the principle of representativeness, the researcher were able to generate accurate and meaningful insights into the technical writing abilities of pre-service EFL teachers. The details regarding the corpus of the study can be seen in Table 1 below.

Table 1. Details of the lesson plan corpus utilized in the present study

Academic Terms	Number of Lesson Plans	Number of Pre-Service Teachers
2018 Spring	270	27
2019 Fall	168	12
2019 Spring	156	13
2020 Fall	160	16
2020 Spring	160	16
2021 Fall	130	13
2021 Spring	154	11
2022 Fall	192	16
2023 Fall	160	16
Eliminated Lesson Plans	150	
Total	1400	140

In EA studies, the literature mentions different types of data: (i) naturally-occurring samples, (ii) clinically elicited samples, and (iii) experimentally-elicited samples (Ellis & Barkhuizen, 2014). While the first one is considered as the most "natural" data, the second and the last ones are regarded as controlled data. In other words, clinically and/or experimentally-elicited samples entails asking to produce specific language forms from language learners. While the participants of the present study were writing their lesson plans, they were not asked to produce certain grammatical structures but rather design a proper English lesson for specific learners groups. Considering this, the data collected for this research can be categorized as naturally-occurring sample since it was "produced in a real-life situation" (Ellis & Barkhuizen, 2014, 23).

For conducting EA studies, Corder (1975) suggests "we need a qualitative linguistic classification of errors, and a quantitative statement of the relative frequency of each type of error" (p.205). For this reason, in the present study, identifying errors entailed qualitative analysis and reporting the results were done in quantitative fashion. For the qualitative part, the corpus data were organized, and a randomly selected lesson plan sample of each participant was analyzed. Coding of the lesson plans was done by the researcher in two rounds for reliability. During the first coding, the researcher identified the errors and the related categories based on the Surface Structure Taxonomy (Dulay, et al. 1982). Once the initial coding was over, the researcher went over the data to check if the errors and the categories matched. The researcher then re-identified the errors in the selected portion of the corpus and coded them into the categories. Finally, the two rounds of coding were compared for reliability (Creswell, 2013). The agreement percentage was calculated to be 87,4%.

RESULTS AND DISCUSSION

Upon examining the corpus of 1400 English lesson plans, the study identified 764 language errors (SD=4.33, M=5.37). The findings of the study revealed that the majority of the errors made by the pre-service EFL teachers were omission errors, which accounted for 45.2% of the total errors. This finding suggests that the teacher candidates were more likely to leave out an item that was necessary for a grammatically correct sentence. This is a significant finding as omission errors are often seen as indicative of gaps in the participants' knowledge of the target language's grammatical rules. Misinformation errors were the second most common type of errors, making up 30% of the total errors. Misinformation errors occur when the participants use an incorrect form of a morpheme or a grammatical structure. These types of errors can often result in miscommunication and misunderstandings. Furthermore, the results showed that the participants made fewer errors in the additions (17.4%) and misordering (7.3%) categories. Additions refer to the presence of an item that should not appear in a grammatically correct sentence, while misordering errors occur when morphemes are placed in the wrong order in a sentence. These types of errors can sometimes be attributed to the participants' lack of vocabulary knowledge or limited exposure to the target language's syntax. A summary of the results can be viewed from Table 2 below.

Table 2. Summary of the results of the present study

Error Category	Number of Errors	Examples From the Data
Omission	346 (45.2%)	*Do you have any question? *I liked it when I listened, also I hope you will like. *Let's listen the song.
Misinformation	229 (30%)	*Look under the passage, there are some sentences. *They have known how to listen critically. *What were Tina doing?
Additions	133 (17.4%)	*The teacher talks about the himself a little bit. *Do we use the first form of the "be"? *They help to people!
Misordering	56 (7.3%)	*What the teacher is saying? *What is the best movie have you ever seen? *Tell me what did you learn in this lesson?
Total	764 (100%)	

Since errors open windows into the learners' interlanguage, when discussing errors made by language learners, it is important to consider interlanguage, which refers to the (i) process of language transfer, (ii) transfer of training, (iii) L2 learning techniques, (iv) L2 communication strategies, and (v) overgeneralization of L2 rules. According to Selinker (1972), these are the five primary processes that contribute to the development of interlanguage. In addition, second language communication strategies cover the ways in which learners interact with others, including native speakers. This underlines the importance of interlanguage in identifying the underlying causes of learners' errors and aiding researchers in fully understanding the structure and operation of the learner's internal linguistic system.

OMISSION

The fact that omission errors were the most common type of error in the lesson plans written by prospective English teachers is worth further examination. According to Dulay et al. (1982), such errors are more likely to occur among beginner language learners. It is interesting to note that despite the participants being senior-year students training to become English language teachers, omission errors still accounted for a significant percentage of errors in their writing. This finding raises questions about the effectiveness of English language instruction at the tertiary level, and whether teacher training programs are adequately preparing future teachers to teach the language.

Furthermore, the high prevalence of omission errors in the lesson plans is also important because the ability to produce accurate and grammatically correct language is essential for effective communication, particularly in an educational setting. As part of their subject matter knowledge, English language teachers are expected to have a high level of proficiency in the language they are teaching (Berry, Depaepe, & van Driel, 2016). Therefore, this result has implications for the quality of education provided by prospective English language teachers, as well as the perception of their competence by their students and colleagues.

It is noteworthy that previous research studies that utilized the same taxonomy for EA have also found omission errors to be the most common type of error for language learners (Suhono, 2016; Yusuf, Mustafa, Iqbal, 2021; Dewi & Fitria Hardini Rangkuti, 2021). Therefore, the present study's findings align with the existing literature on language EA, adding to the body of knowledge on the linguistic challenges faced by language learners, including future English language teachers.

MISINFORMATION

Misinformation category has the second most frequent errors (30%) in this data set. The errors in this category stem from either "miscomprehension" or "misinterpretation" of the target language grammar rules (Dulay et al.,1982, p. 160). It is suggested that in omission category, the learners do not provide the item whereas in misinformation, the learners provide the item, although it is incorrect. There are similarities between the percentages of misinformation errors in this study and those expressed by Suhono (2016) and Ting et al. (2010). However, contrary to the results of Ting et al. (2010)'s study, in the present study misinformation was the second most frequently observed error category. By investigating written errors of college level English language learners, Ting et al (2010) stated that misinformation had the highest number of errors in their data set, followed by omissions. Similarly, El-Mahdy (2023) found misinformation to be most frequently identified error type. On the other hand, both in the present study and in others (Suhono, 2016, Yusuf et al., 2021; Dewi & Fitria Hardini Rangkuti, 2021) misinformation category was the second most frequently observed error group.

The high occurrence of misinformation errors among prospective EFL teachers in the current study is important because it indicates that these learners might be relying on incorrect grammar rules and structures, which can affect their future teaching practices. Vélez-Rendón (2002) argued that teachers' subject matter knowledge, which includes their language proficiency, is essential for effective teaching. Therefore, it is essential to provide these prospective teachers with targeted language instruction and feedback to help them overcome these errors. Misinformation errors can also stem from interference from the learners' first language (L1) grammar. Studies have shown that L1 influence can be a significant factor in learners' language acquisition (Jin, 2009; Spada & Lightbown, 1999). In the present study, the participants were all EFL learners from an expanding circle context, where English is not the primary language of communication. It is possible that the participants' L1 grammar interfered with their acquisition of the target language grammar, leading to the high occurrence of misinformation errors.

ADDITIONS

The additions category of errors, which indicates the presence of an item that must not appear in a well-formed sentence, was the second least frequent category observed in the current study, accounting for 17% of the errors. Dulay et al. (1982) suggest that this type of error may occur at later stages of L2 development, when learners have acquired certain grammatical rules but have not yet fully refined their understanding and use of them. This finding is in line with the results of previous studies that used the same EA taxonomy (Suhono, 2016; Ting et al., 2010; Yusuf et at., 2021; El-Mahdy, 2023; Dewi & Fitria Hardini Rangkuti, 2021), which also reported lower frequencies of addition errors compared to omission and misinformation errors. These results suggest that the participants in the current study may have a relatively strong command of English grammar, but they still struggle with certain aspects of the language. It is important for future EFL teachers to develop a high level of proficiency in the target language in order to effectively teach it to their students.

MISORDERING

Misordering errors, or the incorrect placement of morphemes within a sentence, were the least frequent type of error observed in the current study, accounting for only 7.3% of all errors. As noted by Dulay et al. (1982), misordering errors can occur in both simple and embedded questions and can often be attributed to interference from the learner's first language. This finding is supported by the present study, as the majority of misordering errors observed (N=56) were related to the incorrect formation of interrogative statements. These results are consistent with previous studies that used the Surface Strategy Taxonomy (Suhono, 2016; Ting et al., 2010; Yusuf et al., 2021; El-Mahdy, 2023; Dewi & Fitria Hardini Rangkuti, 2021), which also reported relatively low frequencies of misordering errors. This may suggest that learners are able to develop a better understanding of the grammatical rules related to word order as they progress in their L2 development.

CONCLUSION

In conclusion, EA is a crucial component of second language acquisition research. Researchers can recognize and comprehend the types and underlying reasons of incorrect forms of learner language by classifying them. Based on the results of the present study, informed by both Surface Strategy Taxonomy and EA, efficient language education and curriculum construction are possible. It is possible to encourage more accurate language use by providing corrective feedback and broad context for language users to practice the target language. Errors used to be seen as sins to be avoided at all costs (Brooks, 1964) however, thanks to EA they are now seen as "indicators of learning and guides to teaching" (Dulay et al., 1982, 197). Using error categorization or taxonomies helps researchers to identify the problematic areas in learner interlanguage, yet one cannot attribute "single source to each error that occurs" (Dulay et al., 1982, 197). Error analysis focuses on tangible data, namely actual language use by the learners, and thus determining underlying causes of why errors occur in the first place requires studying internal (e.g., cognitive processes, L1 and L2 structures) and external factors (e.g. language learning environment, learner differences) at work. Accordingly, this study aimed to provide an analysis of future EFL teachers technical writing errors and offer an understanding of the most frequent errors for teacher development.

This study is a significant attempt to provide insights into the context of English language teaching and learning in an EFL setting, specifically at a university where participants are training to become English language teachers. As mentioned earlier, the expanding circle is a unique context for English language learners as compared to the inner and outer circles, due to limited opportunities for receiving target language input and producing output opportunities. Despite these challenges, the participants of this study have taken steps to become proficient in the target language. They have passed nation-wide university entrance and target language exams, completed undergraduate-level courses in the target language, and engaged in a teaching practicum that integrates their theoretical and practical knowledge as teachers. Given the participants' goals of becoming English language teachers, it is vital for them to have a strong command of the language and the ability to write effective lesson plans. Writing lesson plans requires a specific set of skills and knowledge, including the use of specific terminology, grammatical structures, and a clear understanding of the learning outcomes of the lesson. Therefore, the study examined the writing practices of pre-service EFL teachers by analyzing a corpus of English teaching lesson plans written by them. This analysis provides insights into the language errors that pre-service EFL teachers make in their technical writing, which can be addressed to improve their language proficiency and effectiveness as future teachers. Overall, this study is essential in understanding the unique challenges and opportunities present in the expanding circle context and how it affects the language learning and teaching of EFL learners. It highlights the critical role of writing effective lesson plans in English language teaching and the need for pre-service EFL teachers to develop their language proficiency to be effective in their future profession.

Additionally, the study's findings emphasize the importance of EA in improving the quality of language teaching and learning. By identifying the most common types of errors made by pre-service EFL teachers in their technical writing, the study provides insight into areas where language teachers need to focus their attention. Understanding these errors can help teachers design more effective lesson plans that target specific areas of weakness and help students improve their language skills more efficiently. The study's results are also useful for teacher training programs, as they can be used to develop curricula that specifically target the areas where pre-service EFL teachers need to improve their language skills.

IMPLICATIONS

The results of the study can be used for designing and refining undergraduate English teacher education programs. Based on the results of this study, it is observed that future EFL teachers could benefit from receiving more training on the grammar of target language before they graduate. Teacher education programs offered at undergraduate level could include grammar courses, integrate technical writing exercises into the curriculum, and exercise self-correction techniques for prospective English teachers. Additionally, programs that prepare language teachers can gain a lot from making EA a major part of the curriculum. Teacher trainees can learn to recognize and understand errors committed by their own students by being exposed to EA. As a result, they may be able to create teaching methods that are better suited to the needs of their students. Moreover, teaching professionals who have had experience to EA tend to be more introspective. Teachers can evaluate their own teaching strategies and spot areas for development by looking at student errors. This may result in a more engaging and successful teaching method.

The study may also be fruitful for the professional development of in-service English teachers. Based on the results of the study, in-service training programs can incorporate professional English courses for teachers, help them identify and correct their errors through awareness-raising activities, and may offer a network for exchanging feedback on their written work. Error analysis can also aid in increasing teachers' understanding of the difficulties experienced by their students. Teachers can foster greater empathy for their students and improve the learning environment by comprehending the nature of learner errors. Furthermore, curriculum designers may supplement additional courses for improving the technical writing practices of pre-service EFL teachers, or the components of existing courses can be overviewed to include tasks related to the most commonly observed errors.

FUTURE RESEARCH DIRECTIONS

In spite of its limitations, the present research certainly improves our knowledge of the technical writing errors of prospective EFL teachers. Considering the lack of EA studies related with language teacher lesson planning, further studies may contribute to the literature by investigating in-service teachers' technical writing. The current study followed the Surface Structure Taxonomy (Dulay et al., 1982) which proposed four main categories of learner errors. Others could also employ different error categorization (Ellis & Barkhuizen, 2014; James, 2013) to understand the issue from other perspectives. The participants of this study consisted of future EFL teachers in Turkey; therefore, the data set is limited to this particular context.

Additional research is needed to have a more profound understanding of written language errors of pre- and in-service EFL teachers. A greater attention on future and in-service language teachers' written work could produce remarkable findings that account more for teacher subject matter knowledge.

ACKNOWLEDGMENT

This research received no specific grant from any funding agency in the public, commercial, or not-for-profit sectors.

REFERENCES

Al-Sobhi, B. M. S. (2019). The nitty-gritty of language learners' errors—Contrastive Analysis, Error Analysis and Interlanguage. *International Journal of Education and Literacy Studies*, *7*(3), 49–60. doi:10.7575/aiac.ijels.v.7n.3p.49

Almusharraf, N., & Alotaibi, H. (2022). *An error-analysis study from an EFL writing context: Human and automated essay scoring approaches*. Technology, Knowledge, and Learning. doi:10.100710758-022-09592-z

Arıkan, A. & Yılmaz, A. F. (2019). Pre-Service English Language Teachers' Problematic Sounds. *International e-Journal of Educational Studies, 4*(7), 1-26. doi:10.31458/iejes.594715

Atmaca, Ç. (2016, April). Error analysis of Turkish EFL learners a case study. *GlobELT 2016 Conference*. 10.1016/j.sbspro.2016.10.007

Aufa, M., & Syarif, H. (2020). Error analysis on classroom language made by pre-service teachers of English education program UNP. *Journal of English Language Teaching*, *9*(1), 56. doi:10.24036/jelt.v9i1.108151

Berry, A., Depaepe, F., & van Driel, J. (2016). Pedagogical content knowledge in teacher education. In J. Loughran & M. L. Hamilton (Eds.), *International Handbook of Teacher education* (pp. 347–386). Springer. doi:10.1007/978-981-10-0366-0_9

Brooks, N. H. (1964). *Language and language learning; theory and practice*. Harcourt, Brace and World.

Brown, H. D. (2014). *Principles of language learning and teaching: A course in Second language acquisition* (6th ed.). Pearson Education.

Castillo Rodríguez, C., & López Pérez, S. (2019). To be or not to be...? analysis of misuses with be in a learner corpus in an online environment. *Revista De Filología De La Universidad De La Laguna*, (39), 153–170. doi:10.25145/j.refiull.2019.39.06

Chan, A. Y. (2010). Toward a taxonomy of written errors: Investigation into the written errors of Hong Kong Cantonese ESL learners. *TESOL Quarterly*, *44*(2), 295–319. doi:10.5054/tq.2010.219941

Connolly, A. J. (2020). Adjective-noun order: An error analysis of Colombian learners of English. *GIST – Education and Learning Research Journal*, *20*, 231–250. doi:10.26817/16925777.708

Corder, S. P. (1967). The significance of learners' errors. *International Journal of Applied Linguistics*, *5*(4), 161–170.

Corder, S. P. (1973). *Introducing applied linguistics*. Penguin.

Corder, S. P. (1975). Error Analysis, Interlanguage and Second Language Acquisition. *Language Teaching & Linguistics: Abstracts*, *8*(4), 201–218. doi:10.1017/S0261444800002822

Creswell, J. W. (2012). *Educational research: Planning, conducting, and evaluating quantitative and qualitative research* (4th ed.). Pearson.

Creswell, J. W. (2013). *Qualitative inquiry and research design: Choosing among five approaches*. Sage Publications.

Crosthwaite, P. (2018). Does EAP writing instruction reduce L2 errors? evidence from a longitudinal corpus of L2 EAP essays and reports. *International Review of Applied Linguistics in Language Teaching*, *56*(3), 315–343. doi:10.1515/iral-2016-0129

Dewi, R. C., & Fitria Hardini Rangkuti, S. (2021). Grammatical Errors on EFL Students' Conversation Practice: Surface Strategy Taxonomy. *Journal Of Education And Teaching Learning*, *3*(3), 10–17.

Dulay, H., Burt, M., & Krashen, S. (1982). *Language Two*. Oxford University Press.

El-Mahdy, F. M. (2023). An Error analysis of the grammatical errors of Egyptian EFL learners and a suggested program for enhancing their grammatical competence. *Transcultural Journal of Humanities and Social Sciences, 4*(1).

Ellis, R. (1997). *Second language acquisition*. Oxford University Press.

Ellis, R. (2008). *The study of Second language acquisition*. Oxford University Press.

Ellis, R., & Barkhuizen, G. (2014). *Analysing learner language*. Oxford University Press.

Erdoğan, V. (2005). Contribution of error analysis to foreign language teaching. *Mersin Üniversitesi Eğitim Fakültesi Dergisi, 1*(2).

Farrell, T. (2002). Lesson Planning. In J. Richards & W. Renandya (Eds.), Methodology in Language Teaching: An Anthology of Current Practice (pp. 30-39). Cambridge University Press. doi:10.1017/CBO9780511667190.006

Fumero, K., & Wood, C. (2021). Verb errors in 5th-grade English learners' written responses: Relation to writing quality. *Languages (Basel, Switzerland)*, *6*(2), 71. doi:10.3390/languages6020071

Gass, S. M., Behney, J., & Plonsky, L. (2020). *Second language acquisition: An introductory course*. Routledge. doi:10.4324/9781315181752

Graves, S. (2010). Mentoring pre-service teachers: A case study. *Australasian Journal of Early Childhood*, *35*(4), 14–20. doi:10.1177/183693911003500403

Hunt-Gómez, C. I., & Navarro-Pablo, M. (2020). Analysis of pre-service foreign language teachers' incorrect articulations: Frequency, influence on communication, and a specific corrective strategy. *Problems of Education in the 21st Century, 78*(6), 933–947. doi:10.33225/pec/20.78.933

James, C. (2013). *Errors in language learning and use: Exploring error analysis*. Routledge. doi:10.4324/9781315842912

Jin, F. (2009). Third language acquisition of Norwegian objects: Interlanguage transfer or L1 influence. In Y. I. Leung (Ed.), *Third language acquisition and universal grammar* (pp. 144–161). Multilingual Matters. doi:10.21832/9781847691323-010

Lee, J. (2020). An error analysis of L2 writing: How does task complexity affect L2 errors? *The Journal of AsiaTEFL, 17*(3), 954–965. doi:10.18823/asiatefl.2020.17.3.13.954

Lelakova, E., & Toman, M. (2023). A corpus based study of commas use in EFL written performance. *International Journal of Instruction*, *16*(1), 1–20. doi:10.29333/iji.2023.1611a

Longcope, P. (2009). Differences between the EFL and the ESL language learning contexts. *Studies in Language and Culture*, *30*(2), 203–320.

Montrul, S. (2011). Morphological errors in Spanish second language learners and heritage speakers. *Studies in Second Language Acquisition, 33*(2), 163–192. doi:10.1017/S0272263110000720

Najlah, N., & Fatimah, S. (2020). Common Grammatical Errors in Oral Communication Made by Indonesian EFL Learners. *Journal of English Language Teaching, 9*(4), 740–749. https://doi.org/https://doi.org/10.24036/jelt.v9i4.110341

Nikolenko, O., Rebenko, M., & Doronina, N. (2021). Academic literacy: The analysis of first-year Ukrainian University Students' errors in essay writing. *Revista Romaneasca Pentru Educatie Multidimensionala, 13*(1). doi:10.18662/rrem/13.1Sup1/396

Peng, S. (2019). A Study of the Differences between EFL and ESL for English classroom teaching in China. *IRA International Journal of Education and Multidisciplinary Studies, 15*(1), 32–35. doi:10.21013/jems.v15.n1.p4

Purgason, K. B. (2014). Lesson planning in second/foreign language teaching. In M. Celce-Murcia, D. M. Brinton, & M. A. Snow (Eds.), Teaching English as a second or foreign language (pp. 362–379). National Geographic Learning, Heinle Cengage Learning.

Rehman, I., Silpachai, A., Levis, J., Zhao, G., & Gutierrez-Osuna, R. (2020). The English pronunciation of Arabic speakers: A data-driven approach to segmental error identification. *Language Teaching Research, 26*(6), 1055–1081. doi:10.1177/1362168820931888

Richards, J. C., & Schmidt, R. (2010). *Longman Dictionary of Language Teaching and Applied Linguistics*. Longman.

Selinker, L. (1972). Interlanguage. *International Review of Applied Linguistics in Language Teaching, 10*(1-4), 209–231. doi:10.1515/iral.1972.10.1-4.209

Shulman, L. S. (1987). Knowledge and teaching: Foundations of the new reform. *Harvard Educational Review, 57*(1), 1–22. doi:10.17763/haer.57.1.j463w79r56455411

Spada, N., & Lightbown, P. M. (1999). Instruction, first language influence, and developmental readiness in second language acquisition. *Modern Language Journal, 83*(1), 1–22. doi:10.1111/0026-7902.00002

Suhono, S. (2016). Surface strategy taxonomy on the EFL students composition a study of error analysis. *Jurnal Iqra'. Kajian Ilmu Pendidikan, 1*(2), 1–30.

Sun, R. (2017). Error analysis in English writing of Engineering College students based on Pigai.org. *Proceedings of the International Conference on Education Innovation and Social Science (ICEISS 2017)*. 10.2991/iceiss-17.2017.45

Ting, S. H., Mahadhir, M., & Chang, S. L. (2010). Grammatical Errors In Spoken English Of University Students In Oral Communication Course. *GEMA Online™ Journal of Language Studies, 10*, 53–70.

Uzdu Yıldız, F., & Çetin, B. (2020). Errors in written expressions of learners of Turkish as a foreign language: A systematic review. *Journal of Language and Linguistic Studies*, *16*(2), 612–625. doi:10.17263/jlls.759261

Vélez-Rendón, G. (2002). Second language teacher education: A review of the literature. *Foreign Language Annals*, *35*(4), 457–467. doi:10.1111/j.1944-9720.2002.tb01884.x

Velliaris, D. M. (2016). Potato, Pot-Ar-To. Tomato, Tom-Ar-To: Is teacher quality and teaching quality the same thing? In D. Polly & C. L. Martin (Eds.), *Handbook of research on teacher education and professional development* (pp. 337–355). IGI Global.

Widya Hapsari, E. W. (2021). Error Analysis in Writing Descriptive Text of Pre-Service English Teachers at Kutai Kartanegara University. *Intelegensia: Jurnal Pendidikan Dan Pembelajaran*, *5*(1), 1–12.

Yusuf, Y. Q., Mustafa, F., & Iqbal, R. M. (2021). An inquiry into grammatical errors in writing committed by high achieving EFL students. *International Journal of Language Studies, 15*(2).

ADDITIONAL READING

Burns, A., & Richards, J. C. (2012). *The Cambridge guide to pedagogy and practice in second language teaching*. Cambridge University Press. doi:10.1017/9781009024778

Ferris, D. R. (2003). *Response to student writing: Implications for second language students*. Lawrence Erlbaum Associates. doi:10.4324/9781410607201

Gass, S. M., & Selinker, L. (2008). *Second language acquisition: An introductory course* (3rd ed.). Routledge. doi:10.4324/9780203932841

Han, Z. H., & Tarone, E. (Eds.). (2017). *Interlanguage: 40 years later*. John Benjamins Publishing.

Tarone, E., & Swierzbin, B. (2009). *Exploring learner language*. Oxford University Press.

KEY TERMS AND DEFINITIONS

Additions: A category of language errors referring to the redundant structures that should not be present in well-formed utterances. Example from the data set: *Are your microphones **is** on? (Are your microphones on?)

Lesson Planning: A chance for teachers to think ahead, oversee, design, and evaluate their in-class procedures and materials before entering the class.

Misinformation: A category of language errors referring to the miss-used structures. Example from the data set: *The aim is to **make** students feel comfortable and get them ready for the course. (The aim is to **help** students feel comfortable and get them ready for the course.)

Misordering: A category of language errors referring to the use of structures in wrong order. Example from the data set: *After the students answer what are the texts about, the teacher asks them the types of the texts. (After the students answer what the texts **are** about, the teacher asks them the types of the texts.)

Omission: A category of language errors referring to the missing structures that should be present in well-formed utterances. Example from the data set: *Students may confuse about the difference between present simple and present continuous tense. (Students may **be** confuse**d** about the difference between present simple and present continuous tense.)

Practicum: The practice of teacher education in which the prospective teachers are to attend, observe and teach in an assigned public or a private school to gain the required experience for their profession before graduation.

Pre-Service Teacher: An undergraduate student who is studying to obtain a degree in teaching.

Chapter 4
Errors in the Written Productions of Spanish Undergraduate Engineering Students:
A Comparative Analysis

Antonio Daniel Juan Rubio
Universidad de Granada, Spain

EXECUTIVE SUMMARY

Writing is a language skill that for most FL students is difficult to master. Therefore, students often find it difficult to write a specific composition which is not their FL. Especially for undergraduate technical students, this seems to be a challenging task since the presence of difficulties in any FL writing forces them to commit errors in their written productions. Therefore, what this chapter will identify is those errors made by engineering students in a public Spanish Polytechnic University over the last two years. The authors shall attempt to categorise errors based on their source following the comparative taxonomy proposed by Dulay, Burt, and Krashen. So, a total of 60 essays undertaken by these engineering students were selected taking into account their productions in the task assigned during the semester. The main objective is to prove that the results have undoubtedly demonstrated that the majority of errors committed by these engineering students can be classified as belonging to the interlingual category, being the category with the highest number of occurrences.

DOI: 10.4018/978-1-6684-6222-5.ch004

"Errors in the written productions of Spanish undergraduate engineering students: a comparative analysis"

INTRODUCTION

Nowadays, there is a clear tendency to consider the errors made by students in their process of language learning not as a negative aspect of language learning but, rather on the contrary, as a natural step in the development of their language skills. As Corder stated, "the errors that learners make are a major element in the feedback system of the process that we call language teaching and learning" (Corder, 1981, p. 35).

In the past, teachers considered errors committed by students as something undesirable, something to prevent from occurring at all means. But in the last decades, given the different scientific studies published, researchers came to consider errors as evidence for a creative process in language learning.

Error analysis provides us with scientific evidence for the system of language that students are using at any time in the course of development of their studies. Basically, this analysis consists of four main steps: to identify the errors, to explain the errors, to categorise the errors, and to evaluate the errors.

Consequently, the errors which emerge from the students' learning process can clearly tell us what needs to be taught or fostered and how the learning is ongoing at any time in order to take the necessary adjustments. Quoting Saville-Troike "learner errors are windows into the language learner's mind" (Saville-Troike, 2006, p. 39).

Also, Dulay, Burt, and Krashen (1982) believe that an analysis of errors made by language learners can help us understand the process of language learning deeply. Besides, it also helps teachers to decide on those teaching materials that best fit the learning needs of the language learners. Therefore, at universities, students are taught to master, at least up to a certain level, English in its four different skills: listening, speaking, reading, and writing. And among these four skills, writing is undoubtedly the most important productive to be learned. But not only is the most productive of the four skills, but it is also the one that they usually face with certain suspicion and even fear due to their proven inability to express correctly in written English, regardless of the nature of the task assigned. As a result, this turns out to be problematic when it comes to giving feedback.

We should especially bear in mind the fact that we are not working with students doing an English studies degree at university, whose English level could be expected to be higher than average, but a technical one within the engineering field. Consequently, all this background information has its natural reflection on the numerous errors committed by them in the student written output which had to be undertaken and assessed.

In consonance, the research methodology conveyed in this study shall consist of following the comparative analysis of errors, as proposed by Dulay, Burt, and Krashen (1982), in order to categorise the errors made in the written productions of engineering students over the last two academic courses. This analysis will provide us with a clear perspective on which category of errors is the most frequent amongst them out of these four: ambiguous errors, developmental errors, interlingual errors, and other errors.

STATE OF THE ART

Along the present chapter we will delve into an account of the main approaches and researches regarding error analysis. According to Saville-Troike, error analysis is the very first approach to the study of Second Language Acquisition (SLA). This analysis includes an internal focus on "the learner's creative ability to construct the language" (Saville-Troike, 2006, p. 38). Therefore, it means it is the first step we must take in order to grasp the acquisition of a second language.

Error Analysis provides evidence not only for the system of language which a learner is using at any particular point in the course of his or her L2 development, but also for the different learner's strategies in order to discover the language. Therefore, the errors which emerge from that learning process could tell the teacher what really needs to be taught and the researcher how the learning is actually proceeding.

Error Analysis is concerned about the learners' creative skill in constructing the second language which is done by describing and analyzing their errors in producing or employing L2 (Saville-Troike, 2012, p. 40). It is also seen as a study of second language learners' morpheme development. Errors are seen as a way to see the learner's mind as they reflect the point of the learner's second language development and give chances to interpret the meaning of those errors as well as, later on, find a way to reduce them.

As an integral part of contrastive analysis, error analysis was used primarily to help language teachers predict what problems a language learner would have due to the linguistic differences between the learner's native language and the target language. Errors that could not be attributed to language interference were practically ignored and those most frequently focused upon, such as the omission of articles before unique nouns or phonological errors, were so well-known that many teachers found the work of the researchers redundant. Thus, in the beginning, error analysis consisted of little more than generalised collections of 'common' errors and their linguistic classification (Ellis, 1985).

Learner errors could provide us with an understanding of the underlying processes of Second Language Acquisition. And these processes, following Richards (1971), can be inferred from an examination of the learner language protocol, case studies, diary studies, classroom observation or experimental studies.

When applied linguists come to confront errors, practical problems arise as defining errors is not in any sense easier than defining learning. Selinker (1972) simply abandoned the concept of "errors" by viewing the language used by the learners as a whole language system, for which he coined the term 'interlanguage', implying that such language is a continuum on the one end of which there is the mother tongue or any previously acquired language and on the other, there is the target language. The interlanguage shares the features of two social dialects of the languages learnt.

For the sources of this language system Selinker identified four possible areas of transfer that might influence the structure of interlanguage: transfer of one's native language or other languages the learner has already acquired, transfer of training, transfer of communication, and transfer of strategy. Nemser coined the term 'approximative system' for the language that the learner is using, implying that the learner is engaged in a gradual process in the direction of the target language.

According to Corder, the procedure for conducting error analysis basically consists of 3 major stages: "recognition, description, and explanation of errors" (Corder, 1981, p. 126). These basic steps were somehow expanded by White, a few years later, who subsequently elaborated these previous three stages into the following steps: collection of data from the students' compositions, identification of errors (labelling the errors with varying degrees of precision), classification into errors types, statement of relative frequency of error types, identification of the areas of difficulty in the target language, and remedial lesson therapy (1987).

Therefore, quoting Richard "errors are usually classified according to language components (phonological, morphological or syntactic)" (Richard, 1974, p. 45). This classification of errors was mostly followed and adopted in any research whose main goal was the analysis and source of errors.

Later on, both Dulay, Burt, and Krashen in 1982 first and James in 1998 later presented the most useful and commonly used bases for a descriptive classification of errors by describing them using different kinds of taxonomy: linguistic category, surface category, comparative taxonomy (the one selected for this research for practical purposes), and communicative effect.

The first taxonomy, linguistic category, carries out errors in terms of where the error is located in the overall system of the target language based on the linguistic item which is affected by the error (James, 1998). It clearly indicates in which component of language the error is located: phonology (pronunciation), syntax, and morphology (clause, noun phrase, verb phrase, auxiliaries, preposition, adjectives…).

This framework is useful and applicable to handle the errors of relatively advanced learners, which is not our object of study.

The second taxonomy, surface strategy category, was also proposed by James (1998). This classification is based on the ways in which the learner's erroneous version is different from the presumed target version. It highlights the ways in which the surface structures deviate. For example, learners may omit necessary items (omission), or add unnecessary ones (addition), or they may just misform items (misformation) or misorder them (misordering). This classification can give us a clear description about the cognitive processes that underlie the learner's reconstruction of the new language which is being learned, which is not really our scope of interest in this paper. It also explains the fact that learner's errors result from their active way in using the interim principles to produce the target language.

The third taxonomy, comparative taxonomy, which is the one we are basing ourselves on in order to conduct this research, was fully developed by Dulay, Burt, and Krashen in 1982. This is a comparison between the structures of second language errors and certain other types of constructions. So, for practical reasons, this is the taxonomy we will be applying along this study since it constitutes an attempt to compare the errors made with the learners' native language. And these comparisons have resulted into four main error categories within this taxonomy: ambiguous, developmental, interlingual, and other.

Different researchers have consistently found that the great majority of errors in the language output of L2 learners, in this case undergraduate engineering students, is of the developmental type. Although adult learners, as in our specific case, tend to exhibit more mother-tongue influence in their errors than children normally do, adult interlingual errors also occur in relatively small numbers according to such researchers. However, as we will also be demonstrating in this research, in our study the great majority of the errors detected in the 130 essays examined correspond to the interlingual type, being incidentally developmental errors the least frequent.

Developmental errors are errors which are similar to those made by children learning the target language as their first language. For example, if we take the following utterance made by a Spanish child learning English "cat eat it", the omission of both the definite article and the past tense marker may be classified as developmental because they are also found in the speech of children learning English as their first language. According to Dulay, Burt, and Krashen, developmental errors can also be classified into the following types: omissions (Mark no milk), additions (Mark doesn't likes it onions), misformations (He gots a flowers), and misordering (What this is?).

Two considerations underlie the interest in comparing L2 and L1 acquisition errors. The first has to do with facilitating L2 theoretical development. As Dulay, Burt and Krashen (1989) mentioned, if the characteristics common to both L1 and

L2 acquisition could be identified, theoretical inferences that have been drawn from the large pool of L1 research data may be applicable to L2 acquisition theory as well. The second consideration has to do with the role of

the first language when learning a second. Since children acquiring a first language have not experienced learning a previous language, the errors they make cannot possibly be due to any interference from another language. When such errors are made by second language learners, it would be judicious to hypothesize that mental mechanisms underlying general language development come into play, not the rules and structures of the learner's native language (Dulay, Burt and Krashen, 1982).

Interlingual errors are similar in structure to a semantically equivalent phrase or sentence in the learner's native language. For example, the utterance "the woman slender" produced by a Spanish speaker reflects the word order of Spanish adjectival phrases [article + noun + adjective]. To identify an interlingual error, researchers normally translate the grammatical form of the learner's phrase or sentence into the learner's first language so as to see if there exist similarities. According to Ellis (1985), this type of error is interference or transfer, which means that the learner's native language interferes with the learning of the L2, or rather it transfers into the learner's developing L2 system.

To identify an interlingual error, researchers usually translate the grammatical form of the learner's phrase or sentence into the learner's first language to see if similarities exist. Interlingual errors, as defined by Dulay, Burt and Krashen (1989), basically refer to L2 errors that reproduce native language structures, regardless of the internal processes or external conditions that spawned them.

Ambiguous errors could be equally classified as developmental or interlingual errors. This type of error reflects both the learner's L1 and also the type of error in the speech of children acquiring English as their first language. For example, in the utterance "I no have a motorbike", the negative construction reflects the learner's native Spanish and it is also characteristic of the speech of children learning English as their first language.

The ambiguous category is principally important in a comparative taxonomy. Assigning such errors to a separate category safeguards the clarity of the findings resulting from a comparative error analysis and enables researchers to describe clear theoretical inferences from the rest of the data (Dulay, Burt and Krashen,1982).

Other errors make up a sort of grab bag for items that do not fit into any other previous category. For example, if we take the utterance "He do thirsty", the speaker used neither the native Spanish structure (in the use of have for is as in He have hungry), nor an L2 developmental form (such as She hungry), where the auxiliary is omitted altogether.

Such an error would go into the other category, also called "unique errors" by Dulay and Burt (1974), referring to their being unique to L2 learners. These errors are not similar to those children make during the first language development, so they must be unique to second language learners, and since they are not interlingual, at least some of them must be unique reflections of a creative construction.

The fourth and last taxonomy is the communicative effect taxonomy, based on the perspective of their effect on the listener and the reader (Dulay, Burt, and Krashen, 1982). It deals with distinguishing between errors that seem to cause miscommunication and those that do not. Errors that affect the overall organization of the sentence usually hinder communication, while those errors that affect a single element of a sentence usually do not hinder communication. Since we are not focusing on oral communication along the present article, this taxonomy cannot aid us explain the origin and source of errors. And according to Dulay, Burt, and Krashen (1982), this type of taxonomy classifies errors into global errors and local errors.

Global errors are errors that affect the overall sentence organization and they significantly hinder communication. These errors include wrong order of major constituents as in "English language use many people"; missing, wrong or misplaced sentence connectors as in "He will be rich until he marry"; missing cues to signal obligatory exceptions to pervasive syntactic rules as in "The student's proposal looked into the principal"; and regularization of pervasive syntactic rules to exceptions as in "We amused that movie very much".

On the other hand, *local errors* are those that affect single elements in a sentence and that do not usually hinder communication significantly, at least not to a great extent. This type of errors includes errors in noun and verb inflections, articles, auxiliaries, and the formation of quantifiers, among some others.

In addition, we cannot skip the latest study carried out by McDowell and Liardet (2020) who employed an error analysis framework, elaborated with the functional descriptions of Systemic Functional Linguistics (SFL), to investigate error patterns in research article manuscripts written by Japanese materials scientists.

Results highlighted the difficulties that the nominal group constituted for participants, with almost half (47.81%) of the identified errors occurring within complex nominal groups. Further, the analysis revealed that the most dominant error pattern involved errors with articles and plural -s. Findings from the study also informed the design of a pedagogical tool to assist Japanese materials scientists and language specialists alike in identifying and rectifying these errors.

Finally, it is also worthwhile mentioning here several studies carried out by some Spanish university professors such as Roca de Larios, Murphy, and Manchon Ruiz among others. In the last decades, this group of professors have published a number of articles dealing not only with the analysis of errors but also with other aspects regarding their interpretation, strategies and components.

METHODOLOGY

This study was carried out using a quantitative research design which was employed to reveal what and how many errors the students made, finding out the most frequent ones following the comparative taxonomy of errors as proposed by Dulay, Burt, and Krashen in 1982, as it was forestalled in the previous section.

The participants involved were a number of undergraduate engineering students, whose age ranged from 18 to 25 and who were mostly male given the technical nature of the degree, doing their engineering degree on a Spanish Polytechnic University. They were asked to undertake a written task on the characteristics of different energy sources as part of their continuous evaluation process. Along the semester, different topics were covered in the English syllabus, including contents such as the use of alternative fuels, mechanics, electricity, wind turbines, robotics, or materials.

For this particular task, which was part of the first unit of the syllabus, they were asked to create a table of their own with different types of energy sources in which they had to analyse their characteristics considering both advantages and disadvantages. They were also supposed to include a proper reflection of their own at the end of the activity so we could check their grade of acquisition.

Regarding the students' English level, we have to consider the fact that in most cases their contact with the English language was limited to their studies during the compulsory Secondary Education for the last six years. Therefore, they are expected to attain a B2 level for the most part. However, there are some students, a very limited group though, whose level is higher than expected since they have been attending private classes or taking English courses in the last years.

The data attained for this research laid the focus on the numerous errors committed by the students in their written compositions. Henceforth, the information obtained was analysed by searching and classifying the errors found into the corresponding categories as established by Dulay, Burt, and Krashen: ambiguous errors, developmental errors, interlingual errors, and other errors.

Therefore, the instruments used to carry out this research corresponded to the writing productions from the students. At this point, it is also important to remark that, prior to the elaboration of the written task, the participants had received a proper amount of information on the topic. For that reason, they had the necessary input of knowledge in order to be able to write about the topic proposed from a theoretical point of view.

As for the procedure followed in this study, it took place during one semester of two consecutive academic courses. Data collection was carried out after the participants had been asked to write the task at home so it was later handed in to the course instructors. The analysis of the results obtained was checked and supervised by both instructors who were in charge of teaching the course following the comparative taxonomy as it was the only taxonomy considered for the present research according to the objectives of the study.

RESULTS

From the analysis of the data obtained, we can anticipate, in general terms, that there was a total of 405 errors found in the 60 essays corresponding to the task accomplished. And following the comparative taxonomy of errors proposed by Dulay, Burt, and Krashen in 1982, the errors found in this task were classified into the following types: ambiguous errors, developmental errors, interlingual errors, and other errors. For the sake of clarity, we just arranged them in alphabetical order, regardless their frequency as we will be discussing from now on.

The written task students had to undertake consisted of writing about alternative fuels in which they had to create a table with different energy sources, considering their characteristics, followed by a brief discussion of pros and cons. All this was meant to be done using comparative and contrast structures. And this task was delivered after the students had been explained a topic on the different alternative fuels existing at the beginning of the semester.

For this task, a total of 60 compositions were gathered and analysed over the last two academic courses. Having a close look at these 60 essays, in terms of the number of errors detected, in only 5 of them no errors were detected, which represented a scarce 8.3%. And quite expectably, the most represented group was the one composed by compositions having more than 5 errors, with a subtotal of 32 and a 53.4% overall. The rest was made up of essays with 1 or 2 errors, 12 out of 60 and a 20% of the total, and those with 3 or 4 errors, 11 out of 60, representing a 18.3%.

Focusing thus on the 55 compositions with at least one error for this first task, a total of 405 errors were detected overall. Following the comparative taxonomy classification, the most numerous category was that of *interlingual errors* with a subtotal of 149 errors, which represented a 36.8% of the total. And as it will be shown in table 1, interlingual errors were divided into the following types: addition of past tense, omission of preposition, inappropriate noun phrase, verb-number disagreement, inappropriate demonstrative, inappropriate preposition, and inappropriate pronoun.

Table 1. Number and percentage of interlingual errors in task 1

Type of Error	Number of Errors	Percentage
Addition of past tense	6	4%
Omission of preposition	8	5.3%
Inappropriate noun phrase	8	5.3%
Verb-number disagreement	92	61.8%
Inappropriate demonstrative	17	11.4%
Inappropriate preposition	16	10.7%
Inappropriate pronoun	2	1.3%
TOTAL	**149**	**36.8%**

Source: Own elaboration

As table 1 shows, 7 different types of interlingual errors were found in the students' compositions. If we refer to their frequency, the most outstanding type was that of the case of verb-number disagreement with 92 errors out of 149, representing a 61.8% overall. Considering the technical nature of the task, it seems reasonable to figure out this as being the most numerous since students tend to forget to pay attention to the concordance between the subject and the verb in the sentences. An example of this type of interlingual error could be: "In my opinion, fossil fuels has got two main advantages".

Some other recurrent interlingual errors were either the use of an inappropriate demonstrative, with 17 errors detected standing for a 11.4% of the total, or the use of an inappropriate preposition, with a similar figure, 16 errors which made up a 10.7%. An example of the former, an inappropriate demonstrative, could be the following; "In this days solar energy can be very profitable", whilst as an example of the latter, an inappropriate preposition, we could mention: "Electricity depends of the wind". In the first case, students tend to overgeneralise the use of the correct demonstrative without thinking on their number. In the second case, we can observe a direct influence of the mother tongue in the wrong use of prepositions.

Two other types of interlingual errors with the same number of cases identified were the omission of a preposition and the use of an inappropriate noun phrase, with 8 errors each, which represented a 5.3% of the total. An illustrative sample of the omission of a preposition can be the next one: "Forget ___ ethylic alcohol, we make to make particular chemical processes with ethanol". This case also reflects the influence of the mother tongue, since it is not necessary to use a preposition in Spanish in such cases. On the other hand, an example of the use of an inappropriate noun phrase can be the following: "It contributes to rain acid". Again, we can perceive the influence of Spanish, following the natural order in Spanish rather than the correct English one.

The last but one type of interlingual error was the addition of the past tense, whether it being regular or irregular, with just 6 cases spotted, making up a scarce 4% overall. As an example of this type of error we can include this sentence: "Solar cells and panels didn't worked at night". Double marking is a common mistake for Spanish students since they are used to including it twice in their mother tongue.

Finally, the least recurrent type of interlingual error in terms of frequency was the use of an inappropriate pronoun, something which happened just twice in the 55 compositions analysed, which roughly embodied a 1.3% overall. As an example of this type of error we could cite the next one: "... and its only affect us with sounds". At least, it seemed that the pronouns were correctly used along the first task so we cannot get definitive conclusions from this particular type of error.

The second most numerous category corresponded to those errors labelled as *other*, not belonging to any of the other 3 categories, with a total of 110 errors out of 405, which signified a 27.1% of the total. Nonetheless, a mighty justification for such a number may well lie in the fact that spelling errors have been included under this category and, taking again into account the technical nature of the writing, it may seem reasonable to expect such a figure of words incorrectly written since the presence of the rest of types is rather residual in most cases.

Other errors have been divided into the following types: addition of article, addition of plural marker, addition of pronoun, omission of the main verb, omission of progressive form, incorrect spelling, and addition of the subject. In the next table we can check the frequency of each type of these other errors.

Table 2. Number and percentage of other errors in task 1

Type of Error	Number of Errors	Percentage
Addition of article	13	11.8%
Addition of plural marker	4	3.6%
Addition of pronoun	1	0.9%
Omission of main verb	1	0.9%
Omission of progressive form	10	9%
Addition of subject	6	5.4%
Incorrect spelling	75	68.3%
TOTAL	**110**	**27.1%**

Source: Own elaboration

Having a look at the numbers, as we previously anticipated, the most common type of other errors referred to those cases in which technical words had been incorrectly written, with a total of 75 errors found, representing a 68.3% overall. This figure can be justified with the great number of technical words that they needed to use in this first task. And a representative example of this type of error could be the following: "It is a renawble source of energy with no fuel costs".

The next illustrative type of other errors in numbers was the addition of an article, whether it was definite or indefinite, when this was not needed with 13 errors identified which stood for a 11.8% of the total. For instance, we could include the next example: "Another type of the combustibles is biocombustibles". It seemed that they tended to overgeneralise the use of the article, even when it was not required due to the influence of the mothr tongue.

The third relevant type of other errors considering its frequency occurred when the progressive form of the verb (-ing) was omitted, something which happened 10 times, denoting a 9% in total. Under this type of error, we could include the next example: "Wind farms are produce more noise for people living near them". Apparently, the correct use of progressive tenses was not fully attained by them in this initial task though there does not seem to be a plausible justification for such error.

Another significant type of other errors in frequency referred to the incorrect addition of a subject, which was found 6 times altogether, denoting a 5.4% of the total. A clear example of such error can be the following: "Solar energy it's the best renewable energy for me". This mostly happened in instances where the subject was impersonal and they tended to duplicate it, overgeneralising its use.

The last but one type of other errors would then correspond to the incorrect addition of a plural marker. This occurred just 4 times along the task, which represented a 3.6% in general. And as an instance we could include the following: "I think it is not a serious problem as others disadvantages". This type of error clearly reflects the influence of their mother tongue since they use the plural form for all types of words in Spanish.

Finally, the least represented types of other errors were both the addition of the pronoun and the omission of the main verb in a sentence, with just one case each, embodying a scarce 0.9% as a whole. The former type of error, the addition of the pronoun, could be found in the next sample: "The advantages are that natural energies they come from natural resources". Meanwhile, as an example of the latter type, the omission of the main verb, we could include this one: "Water energy ___ not got any fuel costs". Their scarce occurrence prevents us from getting definitive conclusions on their misuse.

The third category following this comparative taxonomy in terms of frequency corresponded to those errors classified as *ambiguous*. Ambiguous errors were counted on a total of 85 errors out of 405, which represented a percentage of a 21%

overall. Under this category, the following 7 types have been identified as it will be shown in table 3: inappropriate third person singular, inappropriate auxiliary, omission of article, omission of subject, inappropriate quantifier, inappropriate past, and omission of object.

Table 3. Number and percentage of ambiguous errors in task 1

Type of Error	Number of Errors	Percentage
Inappropriate third person singular	4	4.7%
Inappropriate auxiliary	2	2.3%
Omission of article	10	11.8%
Omission of subject	57	67.1%
Inappropriate quantifier	6	7%
Inappropriate past	2	2.3%
Omission of object	4	4.7%
TOTAL	**85**	**21%**

Source: Own elaboration

As table 3 shows, the most frequent type of ambiguous error happened when the subject of the sentence was omitted, something which happened 57 times along the task, representing thus an outstanding 67.1% of the total for this category of errors. This can also be considered as natural and, up to a certain point, foreseen since we tend to elicit the subject in cases of impersonal subject in Spanish. An example of this type of error can be the following: "Is a renewable energy resource with no fuel costs".

Another repeated type of ambiguous errors referred to the omission of the article, regardless this was definite or indefinite, with a total of 10 cases detected, which made up a 11.8% overall. It was also frequent to come across an essay in which the article had been omitted as in these examples: "In contrast, it's difficult to produce ___ large amount of electricity" (indefinite article); "From all ___ energy sources we discussed about, I think that solar energy could be the best one" (definite article). There is not a logical justification for such errors since they also need them in their mother tongue.

Significant was also the case of the use of an inappropriate quantifier happening in 6 different occasions, which signified a 7% in total. It was not uncommon to find out that an incorrect quantifier had been used. An example of this type of error could be the following: "... as well as destroying much species due to barrage". The distinction between countable and uncountable nouns was not totally acknowledged by the students in this case.

Other not too frequent ambiguous errors were, on the one hand, the use of an inappropriate third person singular and the omission of the object, on the other. In both cases, 4 errors were distinguished which stood for a 4.7% of the total. An example of the type of ambiguous error using an inappropriate third person singular could be the next one: "Fossil fuels generates a lot of electricity". In the meantime, the following could be included as an example of the omission of the object: "There are different fuels and we can classify ___ according to their specific characteristics". No definite conclusions can be drawn from these errors due to their relatively low occurrence.

And finally, the least numerous category in this task following this comparative taxonomy corresponded to *developmental errors* with 61 errors detected out of 405, which represented a little 15% of the total. Unlike other researchers predicted in previous studies, this category did not stand out for the first rank but, on the contrary, for the last one as we have been demonstrating in this research.

Henceforth, under this category of developmental errors we included the following 6 types: omission of auxiliary, omission of plural marker, omission of third person singular, inappropriate part of speech, incorrect auxiliary, and omission of infinitive marker "to".

Table 4. Number and percentage of developmental errors in task 1

Type of Error	Number of Errors	Percentage
Omission of auxiliary	2	3.2%
Omission of plural marker	17	28%
Omission of third person singular	3	5%
Inappropriate part of speech	32	52.4%
Incorrect auxiliary	3	5%
Omission of infinitive marker "to"	4	6.4%
TOTAL	**61**	**15%**

Source: Own elaboration

As can be seen in the previous table, more than half of the total number of developmental errors corresponded to the type of inappropriate part of speech with 32 errors detected, which denoted a 52.4% overall. This type of developmental error can be explained due to the incorrect formation of long complex sentences using technical vocabulary as they were asked for in this initial task. A clear example of this type of error could be the following: "Water energy there are no fuel costs".

The second most numerous type of developmental errors referred to the omission of a plural marker with a total of 17 errors spotted along the task, representing a 28% of the total. The use of numerous nouns and noun phrases may also justify the presence of such a figure, as in the next case: "Three quarters of the electricity generated in the UK comes from power station fuelled by fossil fuels".

The presence of the rest of developmental errors in this task was almost testimonial as could be seen in table 4, with 2, 3, or 4 errors at the most. Such number of errors, 4, was found in those cases when the infinitive marker "to" had been omitted, which stood for a 6.4% in total. The following could be included as an example of this type of error: "Its infrastructures are very expensive ___ generate water energy". This error reflects the influence of their mother tongue since it is not necessary its use in Spanish.

Three errors were just found in two other types of developmental errors, omission of third person singular and incorrect auxiliary, representing a 5% overall. The use of an incorrect auxiliary when forming negative sentences or questions was scarcely found along the task although we did find the following case: "Solar panels doesn't work at night". And so scarce was the omission of the third person singular as in this sample: "It's difficult to build a wave machine that produce a large amount of electricity".

Finally, the type of developmental errors with the least number of cases was the omission of an auxiliary form, something which happened just twice in the 55 essays examined, which embodied a 3.2% of the total. An example of this type of error would be the following: "Natural gas ___ not contaminate so much". Again, the direct influence of the mother tongue can be clearly perceived in this type of error as its use is not necessary in Spanish.

DISCUSSION OF RESULTS

Language learning, like almost any other kind of human learning, involves committing errors, at least up to a certain point. And recent research in applied linguistics emphasised the significance of students' errors in second language learning. Since then, Error Analysis has become one of the major topics of concern in the field of second language acquisition research and quite a few studies have been released ever since.

Along the present chapter, the main categories of errors have been described and analysed following the classification proposed by Dulay, Burt, and Krashen in 1982 into 4 different taxonomies. This initial approach was continued tracing the sources and types of errors taking into account the comparative analysis taxonomy, one of the 4 different taxonomies proposed by these authors, who divided errors into 4 main categories: ambiguous, developmental, interlingual, and other errors.

Although these researchers, and some others, put forward that the majority of errors committed by adult learns correspond to the developmental category, we have consistently demonstrated along the article that in our specific case, for undergraduate engineering students the most frequent category was interlingual in the task examined.

Let us recall back, at this point, that the main corpus of this research was the analysis of the 60 essays undertaken by undergraduate engineering students throughout the written task over the last two academic courses consisting in the description and explanation of different energy sources.

In terms of the global number of errors spotted, a total of 405 errors were detected along the task assigned, which gave us a result of 6.75 errors on average per essay. And having a close look at the number of errors found in the 60 essays, we made a distinction among those essays having no errors, those containing 1-2 errors, those comprising 3-4 errors, and those with more than 5 errors. Therefore, we found that the great majority of essays contained 5 or more errors, 32 essays, representing a 53.4% overall. The second representative group was that of essays having 1 or 2 errors with a total of 12 essays, standing for a 20% in general. The next type was formed by those essays with 3 or 4 errors, 11 essays, denoting a 18.3% in total. Finally, the least representative group was the one formed by essays with no errors with just 5 essays encountered, meaning a scarce 8.3% of the total. These data can be easily understood in the following figure.

Figure 1. Number of errors in the task
Source: Own elaboration

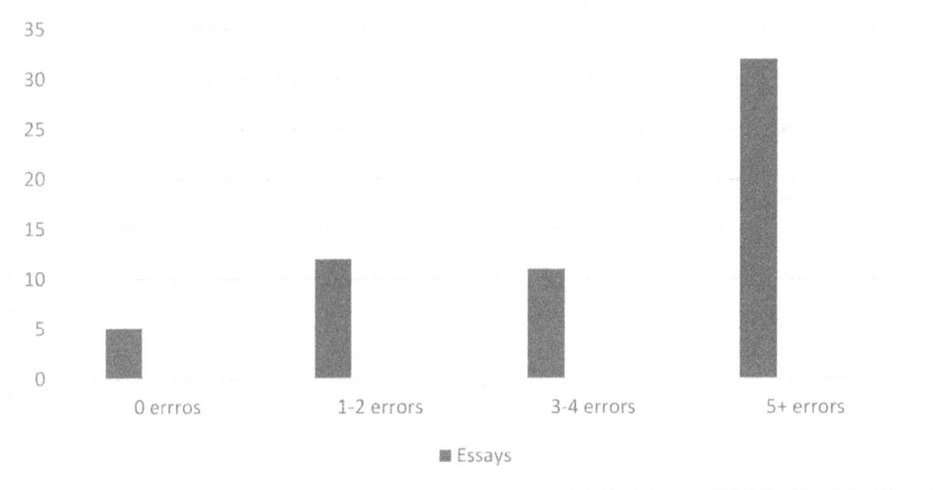

What the previous table clearly shows us is that the highest percentage of essays was concentrated in the group having more than 5 errors. On the contrary, we found the lowest percentage amongst those essays which did not have any error. These data neatly demonstrate the challenge it supposed to these undergraduate students to carry out the written task properly.

More importantly, focusing on the different types of errors in the 60 essays, we proved along the research how the most frequent category of errors was not that of developmental errors but of *interlingual* ones. Contrary to the opinion of some linguists, we have demonstrated that interlingual errors was the category with the highest presence in the task with a total of 149 errors out of 405, which meant a 36.8% overall as can be seen in the figure.

Figure 2. Type of errors in the task
Source: Own elaboration

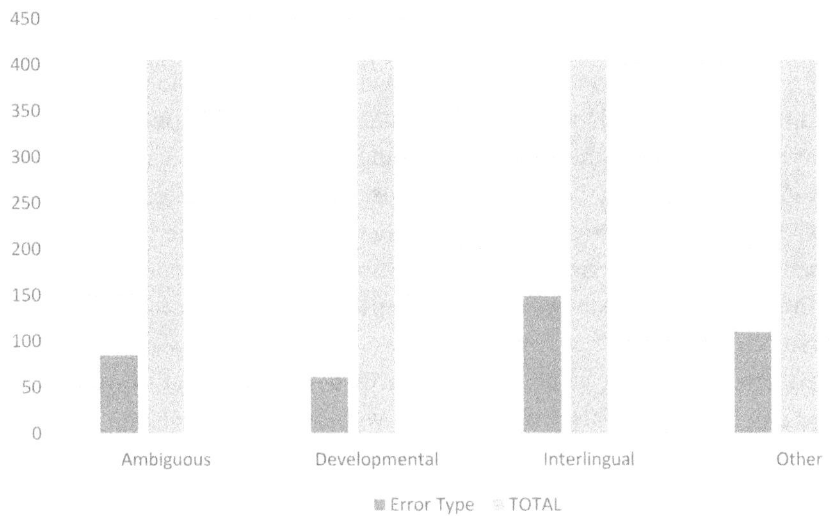

Concentrating now on the type of errors found in the essays, the table revels how the most frequent type of error encountered was that belonging to the interlingual category while the least recurrent error was the one belonging to the developmental category. In between, we can place the other two types of errors, other errors and ambiguous errors.

This might well be explained and justified given the nature of not only the writing examined but also the typology of students looked upon. Perhaps it actually makes a difference whether the students learning English come from a technical degree, undergraduate engineering students in our case, and English is not the goal of their studies but the means, or not in the number and types of errors committed.

Under this category, 8 different types were distinguished: addition of past tense, omission of preposition, addition of preposition, inappropriate noun phrase, verb-number disagreement, inappropriate demonstrative, inappropriate preposition, and inappropriate pronoun. The most recurrent type was the case of verb-number disagreement with 92 errors whilst the least recurrent one was that of the use of an inappropriate pronoun with just 2 error. Other repeated interlingual errors were the use of an inappropriate demonstrative, 17 errors, and the use of an inappropriate preposition with 16 errors.

The second group in terms of numbers was that of *other errors* with a total of 110 out of 405, representing thus a 27.1% overall. However, at this point, it should be clarified that most part of these other errors corresponded to the type of incorrect spelling as we previously mentioned through the article. Taking into account the technical nature of the task, it seemed inevitable to expect such an elevated figure of words incorrectly written. Otherwise, had it not been for this type, other errors would have stood up for the last position in the rank.

Other errors were divided into other 8 following types: omission of indefinite article, addition of article, addition of plural marker, addition of pronoun, omission of main verb, omission of progressive form, incorrect spelling, and addition of subject. And as anticipated, the most numerous type of other errors was that of incorrect spelling with 75 errors due to the scientific nature of the tasks. And the least frequent types of other errors were the addition of a pronoun, and the omission of the main verb with just one error in either case. Besides, the other types were not numerous either.

The next group was then *ambiguous errors*, with 85 errors out of 405, which represented a 21% overall. This group of errors was also divided into some other 8 different types: inappropriate third person singular, inappropriate auxiliary, omission of article, omission of subject, inappropriate quantifier, inappropriate past, omission of object, and inappropriate article. In this case, the most repeated ambiguous error was the case of the omission of the subject with 57 errors. Students faced some problems when they had to include an impersonal subject in the sentences. Meanwhile, the least repeated types were the use of an inappropriate past and the use of an inappropriate auxiliary, with just 2 errors detected in either casa.

Finally, the least represented group, contrary to the opinion of previous researchers, was that of *developmental errors* with a total of 61 errors, standing for a 15.1% as a whole. In this last group, 7 different types of developmental errors were distinguished: omission of an auxiliary, omission of plural marker, omission of regular past tense marker, omission of third person singular, inappropriate part of speech, incorrect auxiliary, and omission of infinitive marker "to". And since students had some problems when constructing sentences in a logical and natural order, the most frequent type of developmental error was the inappropriate part of speech with 32 errors. On the other hand, the omission of the auxiliary was just found twice.

CONCLUSION

Error Analysis is a branch of Applied Linguistics which emerged in the 1960s to reveal that the errors that the learners did were not only due to their mother tongue but rather it also reflected certain universal strategies. This trend supposed a reaction to Contrastive Analysis that considered native language interference as the main source of errors in the field of second language learning. Since then, errors have received much attention from scholars and researchers who tried to analyse different errors made by language learners.

Consequently, procedures have been introduced analysing errors from very different and specific points of view. Many error taxonomies have been based on the linguistic item that is being affected by the error. These linguistic taxonomies classify errors according to the language component and the linguistic constituent that the error affects.

Errors can be classified according to different taxonomies. The current research presented here follows the classification proposed by Dulay, Burt, and Krashen in "Language Two" (1982), being the comparative taxonomy one of the major ones in order to classify errors. They firmly believed that an analysis of errors made by language learners could help us understand the process of language learning deeply. The classification of errors following a comparative taxonomy is grounded on the comparison between the structure of L2 errors and other types of constructions reflecting their native language (Spanish in our case). Therefore, this taxonomy includes four main categories of errors: developmental, interlingual, ambiguous, and other errors.

Therefore, the main objective of this paper was to classify the errors made by undergraduate engineering students in a public Spanish Polytechnic University over the last two academic courses when undertaking a written assignment as part of their continuous evaluation process. These students, whose age ranged between 18 and 25 years old, had an English level equivalent to a B2 as a result of the compulsory secondary education since most of them didn't have any contact with the English language other than the official one. However, a small group of these students had a higher level due to their attendance to private lessons or official English courses.

The task that these students had to undertake consisted of writing an essay about the use of alternative fuels stating both advantages and disadvantages, and finishing with a personal reflection as a way of conclusion. This topic was part of the contents for the English syllabus that these undergraduate students had to take along a semester in their studies. Other topics studied along the course had to deal with mechanics, electricity, wind turbines, materials or robotics.

Focusing now on the written essays that the students had to deliver to this particular task over the last two academic years, a total of 60 essays were gathered and analysed in the present research following the comparative taxonomy by which errors were classified into the following four categories: ambiguous, developmental, interlingual, and other errors.

The results obtained have demonstrated that the most frequent type of error in the case of these undergraduate engineering students was that of interlingual category, reflecting the interference process with their mother tongue (Spanish). The influence and interference of Spanish was reflected in errors committed such as the addition of a past tense, the omission of a preposition, cases of verb-number disagreement, or the use of an inappropriate noun phrase, demonstrative, preposition or pronoun. This type of errors accounted for almost a 37% of the total of errors encountered along the task assigned, which clearly demonstrates their frequency and relevance.

The second most recurrent type of errors was those errors belonging to the other errors category, with a 27% of all the errors encountered in the task. Let us recall back that this category encompassed errors which did not fall into any of the other categories. Henceforth, we could find errors such as the addition of an article, a plural marker, a pronoun, or a subject, the omission of the main verb or a progressive form, and above all, the incorrect spelling of certain words. Given the technical nature of the task and the contents underneath, it was quite reasonable to expect to come across such a number of misspelling errors.

In the third place we found out those errors belonging to the ambiguous category, reflecting their lack of proficiency in the English language. This type of errors represented a 21% of the total of errors we came by in this written assignment. This category included errors such as the use of an inappropriate third person singular, auxiliary, or quantifier, or the omission of an article, a subject or an object. Undoubtedly, their poor English level regarding their writing skills may well be the reason for such a number of ambiguous errors.

Finally, the type of errors with the least presence in the task analysed was those errors which belonged to the developmental category, with just a 15% of the total of errors, reflecting their process of elaboration of proper English written constructions. Under this category we could find errors such as the omission of an auxiliary, a plural marker, the third person singular marker, the infinitive marker "to", or the use of an inappropriate part of speech as a way of mastering the English language.

As a concluding remark, we can end up recalling back the main information gathered along the article. The corpus of this research has been the analysis of errors in 60 essays written by undergraduate engineering students in the last two years following the comparative taxonomy proposed by Dulay, Burt, and Krashen. In these essays, the most recurrent category of errors detected was that of interlingual errors, being the least one developmental errors. In between, we placed other errors in the second place and ambiguous errors in the third.

ACKNOWLEDGMENT

The author would like to thank all the Spanish undergraduate students who granted their permission for making public the errors found out in their written productions.

REFERENCES

Corder, J. P. (1981). *Error Analysis and Interlanguage*. Oxford University Press.

Dulay, H., & Burt, M. (1974). Errors and Strategies in Child Second Language Acquisition. *TESOL Quarterly*, *8*(2), 129–136. doi:10.2307/3585536

Dulay, H., & Burt, M. (1974). Natural sequences in child second language acquisition. *Language Learning*, *24*(1), 23–40. doi:10.1111/j.1467-1770.1974.tb00234.x

Dulay, H., Burt, M., & Krashen, S. (1982). *Language Two*. Oxford University Press.

Ellis, R. (1985). *Understanding Second Language Acquisition*. Progress Press.

Ellis, R. (2003). *The Study of Second Language Acquisition*. Oxford University Press.

George, V. (1972). *Common Errors in Language Learning: Insights from English*: Rowley: Newbury House.

Hyland, K. (2003). *Second Language Writing*. Cambridge University Press. doi:10.1017/CBO9780511667251

James, C. (1998). *Errors in Language Learning and Use*. Longman.

Krashen, S. (1975). An error analysis of an advanced ESL learner: The importance of the monitor. *Working Papers on Bilingualism*, *7*, 125–129.

Kulac, O., Babaoglu, C., & Akman, E. (2022). Public Affairs. *Education + Training*, 21.

Lightbown, P. M., & Spada, N. (2006). *How languages are learned*. Oxford University Press.

Manchon Ruiz, R. M., Murphy, L., & Roca de Larios, J. (2007). Lexical retrieval processes and strategies in second language writing: A synthesis of empirical research. *International Journal of English Studies*, *7*, 149–174.

McDowell, L., & Liardet, C. (2020). Towards specialized language support. An elaborated framework for error analysis. *English for Specific Purposes*, *57*, 16–28. doi:10.1016/j.esp.2019.09.001

Murphy, L., & Roca de Larios, J. (2001). Some steps towards a socio-cognitive interpretation of second language composition processes. *International Journal of English Studies*, *1*, 25–45.

Ramón, N. (2006). Multiple Modification in English and Spanish. *International Journal of Corpus Linguistics*, *11*(4), 463–495. doi:10.1075/ijcl.11.4.06ram

Richards, J. C. (1973). *Error Analysis: Perspectives on Second Language Acquisition*. Longman.

Roca de Larios, J. (1996). Linearization strategies in ELF writing. *Lenguaje y textos*, *8*, 191-208.

Roca de Larios, J., Manchon Ruiz, R. M., & Murphy, L. (2007). Componentes básicos y evolutivos del proceso de formulación en la escritura de textos en lengua materna y lengua extranjera. *Revista Española de Lingüística Aplicada*, *20*, 159–184.

Saville-Troike, M. (2012). *Introducing Second Language Acquisition*. Cambridge University Press. doi:10.1017/CBO9780511888830

Taylor, B. P. (1975). Adult language learning strategies and their pedagogical implications. *TESOL Quarterly*, *9*(4), 391–399. doi:10.2307/3585623

Wang, C., & Winstead, L. (2016). *Handbook of Research on Foreign Language Education in the Digital Age*. IGI Global. doi:10.4018/978-1-5225-0177-0

White, L. (1987). Error Analysis and Error Correction in Adult Learners of English as a Second Language. *Working Papers on Bilingualism*, *13*, 42–58.

Yin, R. K. (2003). *Case study research: Design and methods*. Sage.

Chapter 5
ESP in the Field of Architecture:
Features of Teaching Writing and Some Aspects of Error Analysis

Nadezhda Anatolievna Lebedeva
https://orcid.org/0000-0003-4095-2631
Kherson State Agrarian and Economic University, Ukraine

EXECUTIVE SUMMARY

Architecture is one of the fundamental and promising specialties of a modern agricultural university. It is impossible to imagine the study of world experience in this area without ESP since the advanced design literature is written in English. Writing, as a skill that involves all the senses, is a fundamental means of teaching future architects in a foreign language class. A lot of time is devoted to the study of a professional foreign language in the educational programs of agricultural universities. The purpose of this chapter is a theoretical description of the practical experience of teaching ESP writing to future architects of an agricultural university. The novelty of the work is to describe the Ukrainian experience of teaching writing in a specialized foreign language at the Faculty of Civil Engineering of the Agricultural University based on scientific sources and existing programs.

DOI: 10.4018/978-1-6684-6222-5.ch005

INTRODUCTION

Architecture is one of the fundamental and promising specialities of a modern agricultural university. It is impossible to imagine the study of world experience in this area without ESP since the advanced design literature is written in English. Writing, as a skill that involves all the senses, is a fundamental means of teaching future architects in a foreign language class. A lot of time is devoted to the study of a professional foreign language in the educational programs of agricultural universities. This is due to the demand for specialists in the labour market with knowledge of a foreign language. There is a special need for them. When learning during the educational process, certain difficulties arise, without the elimination of which it is impossible to achieve a sufficiently high level of professionally oriented knowledge of a foreign language. Teaching a foreign language in a Ukrainian non-linguistic university is carried out in the first two years in parallel with other subjects related to the block of social, humanitarian and natural sciences. General professional and special disciplines are studied more deeply in the following undergraduate and graduate courses. A foreign language of a specialized orientation in an agricultural university is studied in the first year of the bachelor's and in the first year of the master's degree, gradually improving and becoming more difficult for students to master. A professional foreign language must be developed by mastering specialized vocabulary, the knowledge of which is unthinkable without such a serious learning tool as writing. It is written work that contributes to memorization and high-quality mastery of professional vocabulary. Future specialists in the field of architecture also acquire the necessary professional skills through writing. When mastering a specialized foreign language through writing, especially typical and practically significant texts on architecture and construction are taken from open sources on the Internet. As a rule, these are dialogues, interviews, newspaper and magazine articles for non-specialists of an agricultural university and articles from scientific journals, encyclopedias, instructions, descriptions of drawings and graphs.

The purpose of this chapter is a theoretical description of the practical experience of teaching ESP Writing to future architects of an agricultural university.

The following tasks are set during the study:

- To highlight the main directions for the development of writing skills for further application in professional activities based on existing programs and scientific sources;
- To find out the mistakes of students in the course of studying specialized texts on architecture and highlight the main features.

The novelty of the work is to describe the Ukrainian experience of teaching writing in a specialized foreign language at the Faculty of Civil Engineering of the Agricultural University based on scientific sources and existing programs.

Methodology. This study is an analysis of the literature and practical results for several years from 2017 to 2023 which made it possible to summarize a significant amount of practical information. During the preparation of the chapter, the following methods were used: analysis, synthesis, the philosophical method of generalization and the method of a systematic approach. The comparative tables are compiled. Also, the errors were analyzed and figures were drawn showing the problems of students. The author of this study compiled a questionnaire to find out opinions about the impact of writing on the process of memorizing specialized vocabulary and mastering grammatical material for further application in the professional work of architects.

BACKGROUND

Based on the works by Kuznetsov, Zaitsev, Stepichev (Kuznetsov et. al., 2016), Proskuryakov (2013), Shor and Galliulin (2021), we can talk about communication between architects at two levels: verbal and non-verbal. That is, in the professional field, architects and builders communicate with the help of plans, drawings, models, computer modelling, and finished built architectural structures. These are independent specific meanings, which are complemented by verbal, that is, the level typical when ordinary people communicate with professionals. "The verbal communication of architects among themselves and with non-professionals can be carried out both orally and in writing in a language called the technical language of architecture. Like any other language of professional communication, the professional language of architects is not a linguistically independent system within the framework of a national language. On the contrary, it arose by differentiation and expansion of the latter. The national, or literary, language creates the lexical basis and grammatical structure for any professional language" (Kuznetsova et. al, 2016, p. 98) Thus, the professional language of architects has the following features, which are shown in Figure 1.

Figure 1. Features of architecture special language (summarized by the author of the chapter)

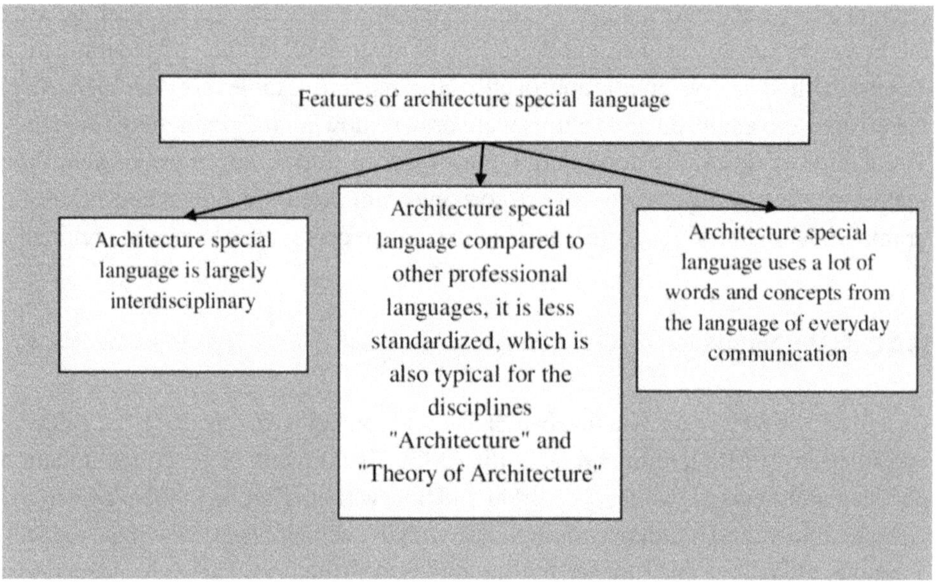

Researchers Kornienko (2013), Zhazanova (2015), Lebedeva (2018), Shor and Gauliullina (2021) note that for students to master a special dictionary and be able to use it, it is necessary to determine the volume of such a dictionary, its organization as a system, and communicate clearly to those who master it. Architects and urban planners often collaborate with engineers. However, the subject matter of architecture is limited to the levels shown in Figure 2.

Figure 2. Levels of the specialty "Architecture" subject study (summarized by the author of the chapter)

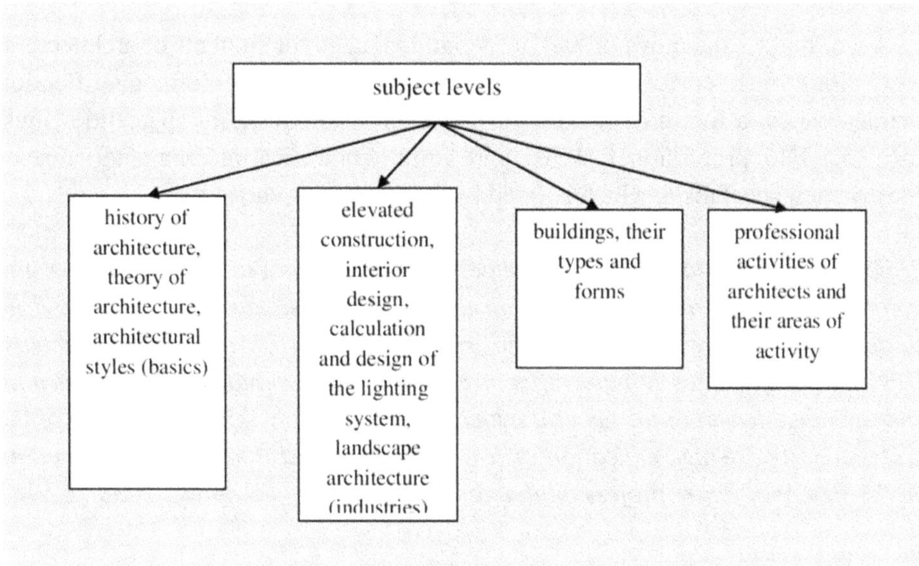

To educate students, along with methodological recommendations developed by universities, teaching aids in a foreign language of professional orientation are widely used. For example, Bezruchko's textbook for architects (2011) is designed to meet the needs of the professional sphere. Some sections help to decide on engineering and construction topics and are useful for students of both architectural and construction specialities. However, it is recommended to work with this textbook for students who already have some base and practice of communication in basic English. The training course helps to improve the speech skills of the English language in the field of construction and architecture. Balanova (2016) regarding the preparation for bachelor's and master's notes the tasks of mastering the professional speech culture, which is achieved by using writing in the classroom.

The researcher notes:

Here we focus on the term "culture" in any interpretation that implies the process and result of continuous improvement. Competences based on the knowledge of the culture of written scientific speech are formed in the study of all disciplines. In the process of studying, students master the glossary on each topic, develop the ability to work with dictionaries, read scientific texts, and gain experience in compiling notes, writing annotations and abstracts, essays, articles and reports, reviews, and reviews. They have skills and experience in editing, compiling a bibliography, etc. (Bazanova, 2016, p. 21)

According to Takanova (2008), a foreign language for students of architectural specialities must correspond to professional skills, and certain qualification characteristics; a foreign language for students of architectural specialities must correspond to professional skills, and certain qualification characteristics of construction specialists, which is fixed by the university curriculum.

Professional-oriented education is understood as education based on taking into account the needs of students in learning a foreign language, dictated by the characteristic features of the profession or speciality, which, in turn, require its study. This is what determines the main difference between professionally-oriented learning and language learning for general educational purposes. Each component of the professionally oriented education system integrates several subsystems depending on the sphere of future professional activity of students. (Takanova, 2008, p. 50)

MAIN FOCUS OF THE CHAPTER

Translation as a Means of a Written Competence Development

Based on the fact that international contacts are expanding, the foreign language of architects is given increasing importance in agricultural universities. One of the most effective means of teaching writing in a specialized foreign language is translation despite the variety of methods, for the preparation of competitive specialists with broad career prospects and advancement opportunities. Possession of written competence in a foreign language contributes to the development of professional qualities of specialists in construction specialities, to an increase in the level of general culture and their outlook, as well as intellect. Translation determines the logic in the construction of thoughts and the presentation of specialized material. Proficiency in a written foreign language will allow graduates to significantly move up the career ladder and expand their educational prospects. Ukrainian higher education is focused on the Bologna system, which is reflected in all programs, which contributes to unification with European standards and determines the further vector of movement in the professional sphere. International, business, economic and cultural ties lead to the need for the ability to write and understand specialized documents and drawings. The transfer of the translation of text features of architectural specialities was considered by Budagyan, Vishnevetskaya, and Romanova (Budagyan et. al., 2015). High-quality writing in a professional foreign language is needed not only for concluding contracts but there is also for research work that involves the translation of foreign literature into the native language and vice versa. At the same time, the translation must be of high quality and sufficiently accurate. Therefore, there is an urgency to learn the written translation of specialized texts on their own, and not in automatic mode. Thus, mental activity is activated, which also positively affects speech competence. Based on the research of Likhachev and Bidenko (2020), Trubin (2016), Kalashnikov, Akai (2016), Levin L. and Levin A. (2019), in the teaching methodology, there are written translation from the native language into a foreign language and vice versa, as well as interpretation. The specificity of an agricultural university is that it is a non-linguistic educational institution, so students have problems with the translation of lexical material, with writing grammatical forms and features from Ukrainian into English. As can be seen from the works of Kuznetsov, Zaitsev, Stepichev (Kuznetsov et.al., 2016), Zhakhanov (2015), Suprunov (2019), Kokorev (2013), students should know the vocabulary of the topic, specialized vocabulary in their speciality, aspectual tense forms of the verb at least in the middle ownership level. Only then will the written decoding process be sufficiently successful. As it is known, written translation allows a student of architectural specialities to see the entire text, reread and revise it. A similar observation is formulated regarding students of technical universities Likhachev and Bidenko (2020). The researchers write the following:

When translating a text, a student memorizes newly translated words, thereby enriching his vocabulary, but in any case, this is only a passive mastery of a foreign language, the development of reproductive thinking. When translating, it is important to take into account its main characteristics - accuracy, transparency and understanding of the text. If these aspects are available, then the translation is of high quality. Moreover, it is important to realize that the translated text must be coherent, consistent with the style, with the right number of synonyms, antonyms and other stylistic means. (Likhacheva & Bidenko, 2020, p. 72)

Likhacheva and Bidenko (2020) also say that it is desirable for a student to feel the content of the translated text and convey the atmosphere, the appropriate time, and expressiveness, in general, which is impossible without the knowledge of the language, speech and linguistic and cultural competencies. Thus, in obtaining the skill of professional writing using translation with students of architectural specialities in an agricultural university, a high-quality translation is needed, which is shown by the teacher, focusing on the presence of possible errors. For students in this case, it is necessary to analyze the specified material and draw conclusions "to avoid such mistakes in the future. It is possible to involve specialized texts, as well as material on business communication and a common language. Translations provided by students must be checked by the teacher, analyzed and evaluated accordingly. It should be noted that teaching translation skills is inseparable from mastering the skills of language, speech and linguistic and regional competencies" (Likhacheva and Bidenko, 2020, p. 73). For students of an agricultural university of construction specialities, it is important to understand that translation is a means of mastering writing, along with other linguistic and communicative competencies, which expands the possibilities for effectively solving professional problems. High-quality professional writing in a foreign language helps to increase competitiveness in the labour market. Written skills allow access to new information and educational resources, which will also expand the qualifications of specialists in architectural specialities in agricultural institutions of higher education. In the classes with students of the Faculty of Civil Engineering of the speciality "Architecture" and "Landscape Architecture", the author of this chapter mainly used the translation of a Ukrainian text into a foreign one, since it is also both a means of teaching a foreign language and its purpose. A large number of tasks in modern foreign language textbooks are devoted to the translation of text or sentences. Many assignments are difficult for students to complete without prior translation, even if the translation itself is not part of the assignment. The perception of a foreign language text should be interpreted as the main means of developing an understanding of specialized vocabulary in architecture. An interesting point of view is expressed by Trubina (2016) regarding students in educational schools. For this study, this is also relevant, since the level of knowledge of a foreign language by students who entered the agricultural university after school is quite low, which entails an additional burden in the process of teaching a specialized foreign language. Teaching how to correctly translate foreign-language texts in the architectural speciality into the native language is one of the important tasks for a teacher. A particular difficulty is precisely written translation since not all constructions of one language are easily translated using another language. According to Trubin (2016), there was no need for translation training for a long time. It was believed that by translating texts, students learn to translate. Now, when considering the requirements for the results of learning a foreign language, students are required to be able to translate simple texts (Trubina, 2016, p. 172). Nevertheless, as the practice has shown, many students experience difficulties even when translating individual sentences.

Written translation is a complex type of speech activity since it is both a receptive and a productive type. We perceive speech through listening or reading, and then generate speech through writing or speaking as a result of the completion of the translation. Translation performs a communicative function, therefore it is a speech activity, with the help of which an act of communication is carried out. In the process of translation activity, the quality of proficiency in all other types of speech activity (listening, reading, writing, speaking) improves. (Trubina, 2016, p. 172)

In written translation, automatic writing skills intersect with creative thinking, however, an important principle of translating a special foreign language text into Ukrainian is that not only words and grammatical structures are translated, but thoughts along with the content of the original. Students are invited to translate both from a foreign language into their native language, and vice versa. Here the technical writing of specialization in a foreign language develops. However, even in this case, translation continues to cause difficulties, since not all constructions of the Ukrainian language are easily transmitted using a foreign one. The complexity of written translation lies in the fact that a student needs to choose the right terms, and grammatical constructions so that the translated text sounds coherent and properly organized. As the classes show, it is difficult for students to complete such tasks without a Google translator. To understand the importance of being able to translate fluently into a foreign language, the author of this chapter offered the translation of instructions, and business letters, both formal and informal. The text for the development of such skills should illustrate the situation. To develop cognitive competence, it's advisable to use specialized texts and to expand personal competence. It is possible to use texts related to the personal qualities of a person. Regarding the professional and technological competence of the architectural orientation, it develops with the help of "the translation activity itself, which is an ordered algorithm of actions" (Trubina, 2016, p. 173). Written translation of a foreign text into the native language fully complies with the requirements of higher education programs, such as Petrosyan, 2019), Zhukova and Burdygo (2019), Lebed and Lebedeva (2018), Shishkina (2020), which contributes to the implementation of students' language competencies. The translation is assessed as an important component of foreign language classes. If the translation is not given in the classroom, then when reading a scientific article, there will be no complete understanding, which will entail a greater number of errors. When using written translation as a means of teaching a foreign language, students of architecture specialities of an agricultural university systematically accumulate knowledge and skills, which in the future will lead to the fact that they will be able to freely perceive the content of a special text on architecture in a foreign language simultaneously with visual perception. According to Uvarov and Erin (2015), an interdisciplinary approach to the organization of the educational process at universities, embodied in various didactic forms, makes it possible to fulfil the requirements of innovative programs.

Figure 3. Main types of translation (summarized by the author of the chapter)

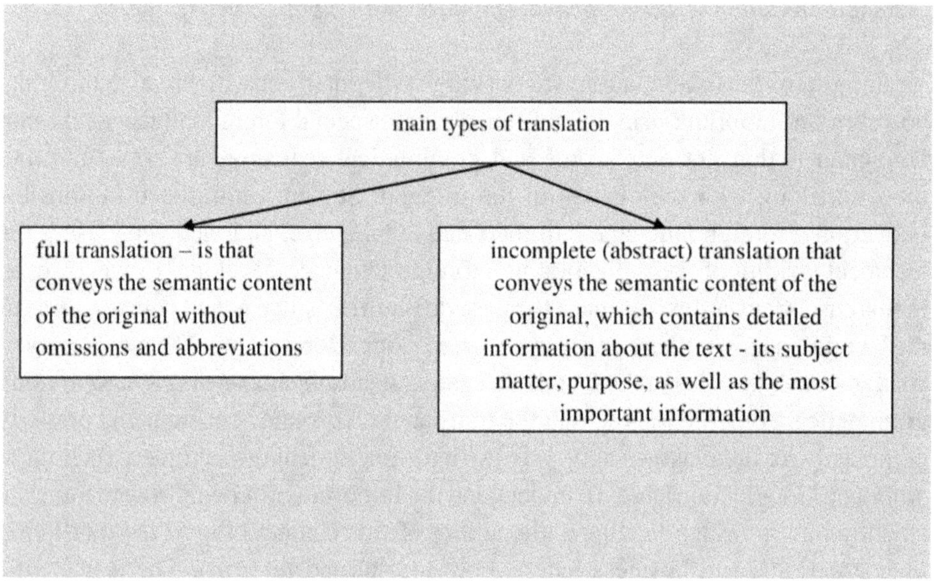

Because the lesson at the university is limited in time, it is more reasonable to teach the written translation of foreign texts into the native language as part of additional classes. But, since written translation contributes to the improvement of a foreign language, it is advisable to introduce elements of written translation in the planned classes, that is, to ask students to translate texts on special architectural topics.

Figure 4. Several types of translation exercises (summarized by the author of the chapter)

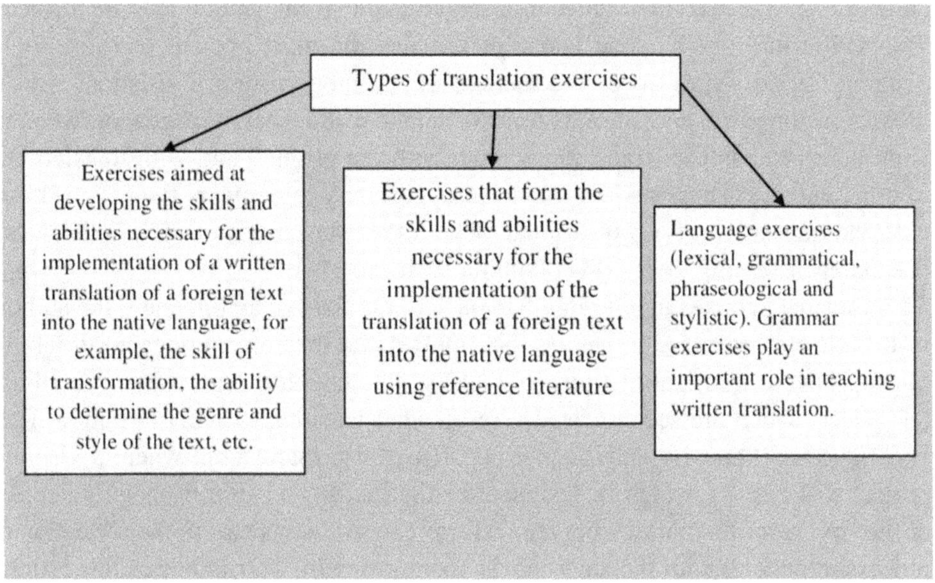

Summarizing the works by Budagyan, Vishnevetskaya, and Romanova (Budagyan et. al.,2015), Proskuryakov (2013), Kornienko (2008), Kuznetsov, Zaitsev, Stepichev (2016), Zhakhanov (2015), Suprunov (2019), Shevchenko and Mitchel (2013), Trubina (2016)), Kalashnikova and Akai (2016), Takanova (2008), Kokoreva (2013), Pakhomkina (2013), the author of this chapter can say that with the help of written language translation exercises, students of architectural specialities of the Agricultural University learn to understand the difference in the grammatical structure of a foreign language and Ukrainian, they learn the skill of conveying the same thought in their speciality with the help of different grammatical structures. Frequent areas of difficulty in the translation of grammatical constructions include the translation of indefinitely personal and impersonal sentences, participles, and the transfer of temporal relations between sentences. The technique of teaching a foreign language of a narrow speciality, used in architectural specialities, offers the vocabulary of the construction field, based on which the grammatical structures of a written text are mastered. With this technique, the skill of transferring any word with an unambiguous equivalent is formed, which may not be suitable for translating a text on a given topic. In this technique, much attention is paid to terms. When teaching writing in a foreign language of architectural orientation, it is advisable to search for variability, when students are offered the absence of an unambiguous correspondence between form and content, that is, the same semantic content can be expressed in different ways. So, for example, it is proposed to find matches for a certain word and write several sentences using synonyms. Thus, the skill and ability to analyze texts of architectural subjects and invent similar ones are formed. Translation into one's native language remains the most popular learning tool. Trubina (2016) writes about the method of teaching written translation, which uses the comparison and analysis of texts in native and foreign languages, when it is proposed to compare translations, both with the original and with translations of other authors. Thus, students have the opportunity to compare their translation with the translation of a professional or with the translations of their colleagues, adequately assessing their skills (Trubina, 2016, p. 174).

In practical classes in a foreign language in the field of architecture, the author of this chapter practised a method that implies that the student does not think to himself how to translate this text, but speaks his thoughts aloud while recording his speech, which showed an effective result when teaching written translation. The experience of Abramov and Ananiev (2010) is also interesting, where a writing project was created, which is designed for the fourth year of a bachelor's degree or the first year of a master's degree and was carried out as part of the educational process in parallel with the study of the main program. The author of this article used some aspects of studying a specialized foreign language for students of a master's program in the speciality "Architecture". Abramov and Ananiev's (2010) initial experience included several stages: 1) a comparative analysis of the styles of written scientific text in English and native languages; 2) the creation of a scientific article in the native language based on the material of the student's term paper in the main speciality; 3) adaptation of this article, taking into account the requirements for English-language scientific texts; 4) translation of an adapted article into English with a pre-translation analysis and a detailed commentary on translation transformations; 5) design and presentation of written work, reflecting all stages of the project. A comparative analysis of the styles of written scientific text in English and native languages, even though it is not a written task, is an important stage, since its results determine the quality of the final text. In most cases, students create a reference text in their native language and then mechanically translate it. At the same time, they do not take into account the fact that the laws on which the Ukrainian text is built do not apply to the text in a foreign language. Errors in translation into a foreign language are superimposed on the shortcomings of the source text. As a result, there is a complete loss of meaning (Abramova and Ananyeva, 2010, p. 96).

To avoid mistakes, a thorough study of the creation of written scientific texts in two languages is suggested. Here, the logical structure, style, and grammar are taken into account, special attention is paid to specialized vocabulary. Thus, students learn the skill of creating a full-fledged scientific article that will represent the results of their research in the future and meet the requirements for scientific papers around the world. A feature of such a written project is that it:

Allows students to develop not only the skills of academic writing but also the basic skills of professionally oriented translation, which can become an important competitive advantage for a modern graduate. Students should be aware that in our time it is written texts (letters, applications for grants, essays for admission to foreign universities, resumes, etc.) that perform a contact-establishing function, make it possible to form an impression of a potential interlocutor or partner and determine the nature of subsequent communication. (Abramova and Ananyeva, 2010, p. 97)

Figure 5. Assistance in the process of completing academic writing assignments (summarized by the author of the chapter)

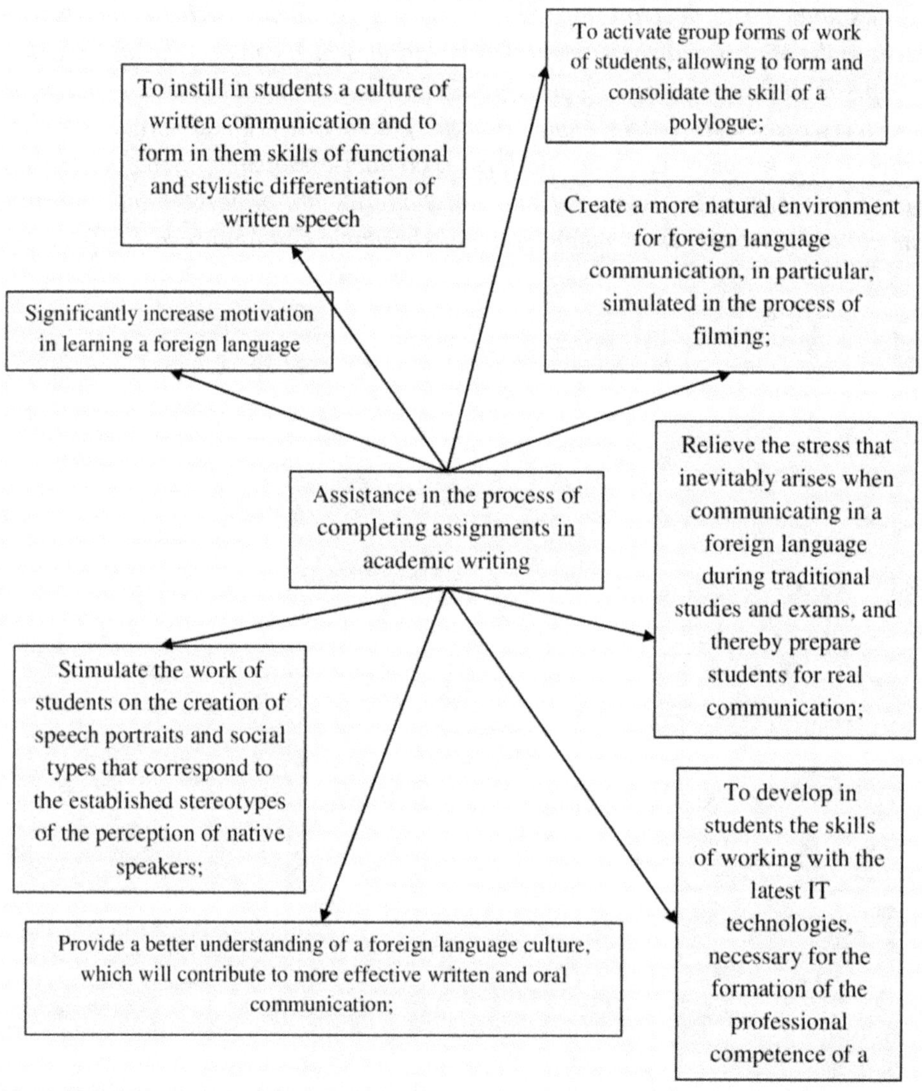

Based on the foregoing, we come to the understanding that teaching writing in a foreign language of a specialized orientation to students of agricultural universities in construction specialities must necessarily include a written translation.

Dictation as an Effective Means of Memorizing Architectural Lexical Units

Dictation has always been and remains one of the most effective means of teaching language competencies. It is used both in the classroom for teaching the native Ukrainian language, as well as a foreign one. Antonova (2017) notes that dictation is one of the oldest means used in teaching writing in native and foreign languages. "The prehistory of dictation takes us back to the time of the existence of scriptoria, workshops for copying manuscripts, where calligraphers wrote books under dictation. In teaching the native language in the Middle Ages, in the absence of books, dictation was practically the only way to convey lesson material, while mistakes in writing were not paid attention to, since the main thing for students was to understand and write down what was dictated" (Antonova, 2017, p. 164). Dictation in the methodology of teaching a foreign language of a specialized orientation is treated differently, since preparation for writing a dictation requires a lot of time from the student, creating difficulties in completing assignments in other subjects. In the practice of modern teaching a foreign language to students of architectural specialities of agrarian universities, dictation is a popular tool even if you are sure that this type of teaching writing can be considered old-fashioned "because it is the simplest form of teaching, involving a" linear "presentation of material, when dictation (input) is followed by comprehension (comprehension) and recording (output)" (Antonova, 2017, p. 164). In class, students record exactly what they have heard, and what is written is the same as the original text. Dictation is a means of controlling acquired knowledge, especially architectural terms. In her scientific research, Antonova (2017) showed that dictations are different, that it is possible to combine their various forms, and that certain transformations of the source text are made, aimed at its restoration, reconstruction, expansion, narrowing, and modification. The scientist also shows that "dictations can be an extremely exciting type of learning activity, being not only a way to test literacy but also a means of stimulating speech and thinking activity, the need to fix one's thoughts in writing" (Antonova, 2017, p. 165).

Baranova and Ignatenko (2016) emphasize that dictations are an integral part of a foreign language lesson and are most often used to control spelling skills. Their work also refers to the fact that dictations are used rather than dictations.

These concepts can sometimes be used as synonyms, but it is in the methodology of teaching foreign languages that they must be distinguished. Dictation is any kind of voice recording to test only spelling or punctuation skills. A dictation is understood as a certain form of recording a sound speech, which involves certain transformations of the original text (reconstruction, modifications, narrowing, expansion). Thus, dictation includes working with the text, dictation involves only writing down the text (words, phrases) without making changes. (Baranova & Ignatenko, 2016, p. 2)

Olifirenko (2017) described her practical result of the organization of the educational process in the study of English in a technical university and its applicability with efficiency, which was confirmed by an increase in students' interest in learning the language and improved academic performance based on surveys and statistical analysis. When modifying the ESP methods, the researcher took into account the industry focus and relevant didactic materials. During the preparation of students, the initial level of language understanding and its characteristics were also assessed. Dictation, generally speaking, is one of the easiest forms of learning to write. This is reading the text aloud in such a way that the student can write down what is read to him word for word, after which the written text is checked and the errors are corrected. In the practice of teaching students in the speciality "Architecture", the author of this study used a specialized dictation.

Dictation based on the book Chudley and Greeno (2006).

Storage Considerations – amount and types of material to be stored, security and weather protection requirements, allocation of adequate areas for storing materials and allocating adequate working space around storage areas as required, sitting of storage areas to reduce double handling to a minimum without impeding the general site circulation and/or works in progress. (p. 81)

Accommodation Considerations – number and type of site staff anticipated, calculate size and select units of accommodation and check to ensure compliance with the minimum requirements of the Construction (Health, Safety and Welfare) Regulations 1996, select sitting for offices to give easy and quick access for visitors but at the same time giving a reasonable view of the site, select sitting for mess room and toilets to reduce walking time to a minimum without impeding the general site circulation and/or works in progress.

Temporary Services Considerations – what, when and where are they required? Possibility of having permanent services installed at an early stage and making temporary connections for site use during the construction period, coordination with the various service undertakings is essential.

Plant Considerations – what plant, when and where is it required? Static or mobile plant? If static select the most appropriate position and provide any necessary hard standing, if mobile check on circulation routes for optimum efficiency and suitability, provision of space and hard standing for onsite plant maintenance if required.

Fencing and Hoarding Considerations – what is mandatory and what is desirable? Local vandalism record, type or types of fence and/or hoarding required, possibility of using fencing which is part of the contract by erecting this at an early stage in the contract.

Safety and Health Considerations – check to ensure that all the above conclusions from the considerations comply with the minimum requirements set out in the various Construction Regulations and in the Health and Safety at Work. (Chudley & Greeno, 2006, p. 81)

As can be seen from the text above, active specialized vocabulary and characteristic grammatical structures are consolidated through written dictation. However, Tsarevskaya and Volodina (2016) are sure that dictation has countless variations that can make the learning process more interesting and personally oriented for learners.

He practices primarily listening and writing skills. And it can also be argued that dictation improves the practical skills of vocabulary, syntax, grammar, as well as reading at the moment when the author analyzes his work. Thus, it can be concluded that dictation trains some practical skills of the language of all speech activity aspects. (Tsarevskaya Volodina, 2016, p. 153)

Tsarevskaya and Volodina (2016) described the use of such a type of dictation as "dictogloss" ("grammar dictation"), which, when performed, must first listen to a text of different length and complexity (depending on the language level of the group), without fixing it on paper. The text is read two or three times, after which it is necessary to write what was heard. "What matters in this exercise is not the fact that they repeat the original sentences word for word, but that they make up sentences that accurately reflect the meaning and structure of the original texts. One of the ways to perform this type of activity is to discuss the ideas of students in microgroups for joint compilation of a version of the original text" (Tsarevskaya Volodina, 2016, p. 154).

Figure 6. Main reasons for using dictation in teaching a foreign language (summarized by the author of the chapter)

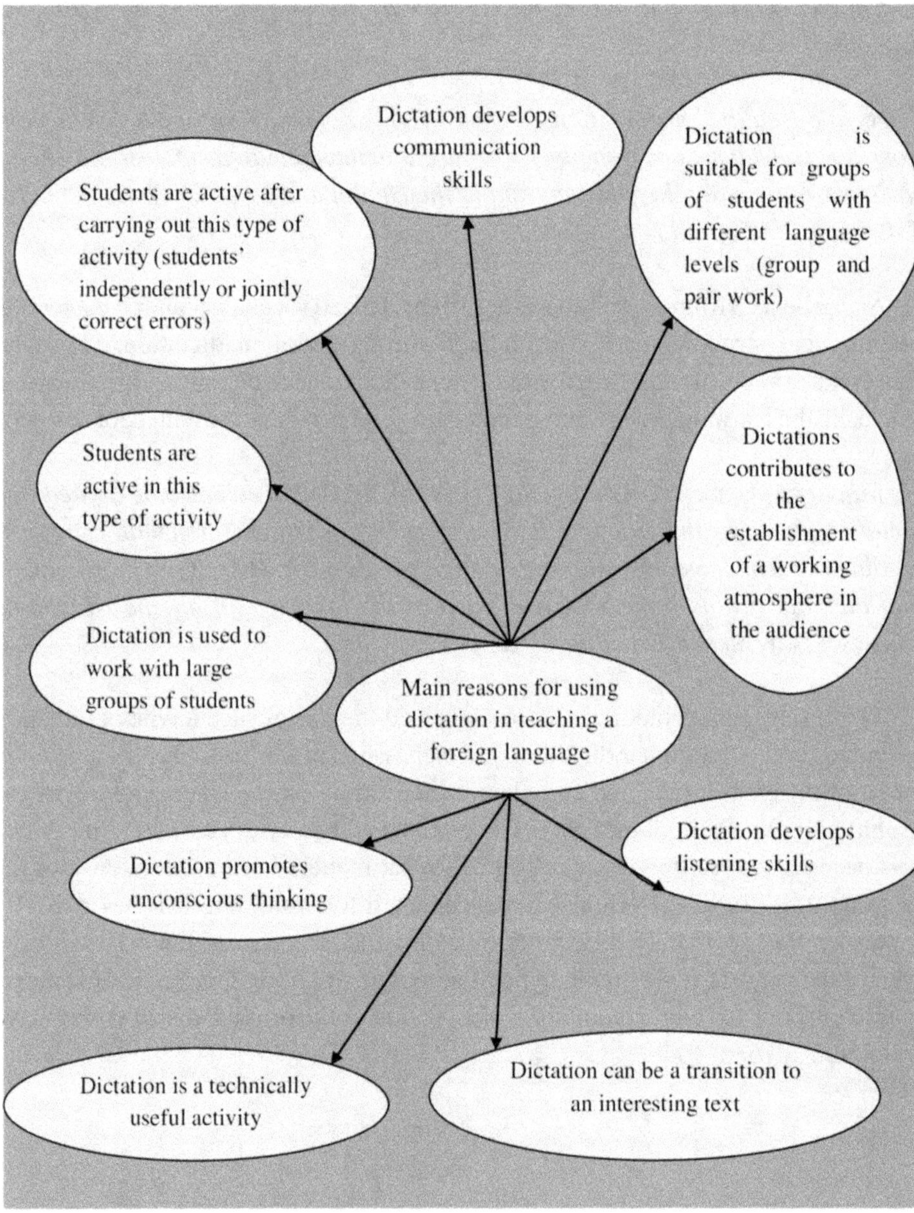

The researchers also write about 4 types of dictations, which in our case can also be used in teaching writing to students of architectural specialities of agricultural universities. Phonemic, dictation consists of the presentation by the teacher of individual sounds of the language to students for writing a transcription. "This type of dictation can be useful for increasing the student's ability to recognize the sounds of the language and their contrasts, thereby ensuring their accurate reproduction," they note (Tsarevskaya Volodina, 2016, p. 154). Although this type of dictation is applicable mainly at the initial stage, it can be quite effective for students with poor pre-university training in a foreign language. The next type used by teachers is text-phonemic dictation. It is a more extended version of the first type of dictation. It is supposed to read a passage of text by a teacher for students to design a phonetic transcription of the text in their speciality. Spelling dictation is the dictation of individual terms in isolation. For example, we used the vocabulary of the textbook "Professional English in use: Architecture. Part 1» by Konovalova, Marsheva, and Mubarakshina (2014).

Theme: GOTHIC STYLE
VOCABULARY
abbey – аббатство parchment – пергамент
castle – замок pattern – узор, шаблон
to check – проверять to prepare – готовить
church – церковь preserve – сберегать
to cut – вырезать quarry – карьер
device – устройство record – запись
to devise – придумывать scaffolding – строительные леса
to draw – чертить, рисовать to shape – придавать форму
to hoist – поднимать sheet – лист
job – работа skill – мастерство
ladder – лестница stonemason – каменщик
medieval – средневековый tool – орудие
molding – отливка, формовка to work out – разрабатывать план

The last type of dictation is the text-spelling or classical type, which is familiar to all foreign language teachers. This type of dictation is aimed at improving the understanding of the text by students and shows gaps in knowledge of grammar. Such a dictation enables teachers to analyze these mistakes and draw students' attention to them (Tsarevskaya & Volodina, 2016, p. 154). It is necessary to select a special text taking into account the modern source, and the standard academic language and taking into account the fact that one of the goals of the dictation is the practice of semantic understanding of a coherent text. For students with a low level of preparation in a foreign language, it is recommended to choose coherent texts with elementary phrases that they already know. At an advanced level, dictations should build on architectural material that students have read, but with more extended content. In both cases, dictations help to improve and consolidate knowledge of basic grammatical and semantic structures, as well as the special vocabulary of architects. At the stage of preparation for writing texts in the speciality, the goal of dictations becomes more complicated. It is necessary to force students to learn to hear and listen to known and unknown facts. A useful source for the selection of material at all levels is the books "Barry's introduction to construction of buildings" (Emmit, Gorse, 2005), "And Building Construction Handbook" (Chudley, Greeno, 2006). Using these sources, the wrong choice of texts unknown to students is excluded. It is best to conduct a dictation at the beginning of the lesson. The text on architecture is read once at normal speed. During the first reading of the text, students only listen. While reading a speciality text a second time at a slower speed. Then the students begin to draft notes. The teacher is encouraged to stop after each paragraph and talk about punctuation to the students. When reading the special text for the third time at normal speed, students check their spelling and make corrections. After the end of the dictation, students are given the source and they correct their mistakes in the text. You can invite students to check each other. Periodically during the semester, the teacher himself needs to check notebooks to correct errors. Evaluation will require attention to separating comprehension and spelling errors. Students of architectural specialities of the agricultural university make comprehension errors, including phonological and grammatical errors. After checking dictations, objective information about the strengths and weaknesses of the students comes and common mistakes can be discussed in the next lesson.

Dictation provides attentive listening, it trains students to recognize sounds, helps to fix punctuation, and enables students to learn how to translate oral sounds into written signs, which helps develop oral understanding, as well as self-esteem. (Tsarevskaya, Volodina, 2016, p. 154)

From all that is written above, it becomes obvious that dictation is a useful and flexible type of activity in teaching a foreign language with an architectural orientation. You can use it when introducing a new structure, repeating and consolidating specialized vocabulary on architecture, and consolidating grammar. Dictation is an important and effective means of teaching writing in a foreign language, as well as a knowledge control tool.

Writing Annotations and Summarizing

Another essential aspect of mastering writing in a specialized foreign language is writing annotations and summarizing. Students of architectural specialities of agricultural universities learn to annotate and abstract specialized materials in a foreign language and master the appropriate terminology, as well as grammatical and lexical means characteristic of foreign documents and scientific correspondence. Shevchenko and Mitchell note that in the course of studying a special course, students receive theoretical knowledge and acquire practical skills in working with documents and deciphering drawings. With the help of abstract writing and summarizing, students develop and improve basic special language skills and skills in the field of professional activity (Shevchenko and Mitchell, 2013, p. 128).

CONCLUSION

After analyzing scientific sources and summarizing her own experience in teaching a written foreign language of a specialized orientation, the author of this chapter concluded that one of the promising areas for the development of higher education in architecture and construction. Almost every institution of higher education in these areas has a training program that includes ESP. Although architects can communicate professionally through drawings and plans, ESP as a discipline does not lose its relevance, on the contrary, it creates a solid basis for communicating internationally through specialized vocabulary and grammatical structures. Undergraduate and graduate students master the culture of professional speech with the help of writing. For honing speech skills and lexical professional skills, the most common means are translation, dictation and annotation. Possession of written competence in a foreign language contributes to the development of professional qualities of specialists in construction specialities, to an increase in the level of general culture and their outlook, as well as intellect. Proficiency in a written foreign language will allow them to move up the career ladder in the future. Being engaged in written translation, new words are memorized and vocabulary is enriched. For students of an agricultural university of construction specialities, it is important to understand that translation is a means of mastering writing, along with other linguistic and communicative competencies. Writing skills allow access to new information and educational resources. In the classes with students of the Faculty of Civil Engineering of the speciality “Architecture” and “Landscape Architecture”, the author of this chapter mainly used the translation of a foreign text into Ukrainian and vice versa.

In translation, automatic writing skills intersect with creative thinking. Also, when translating, a technical letter of specialization in a foreign language is developed. However, translation continues to be difficult, resulting in speech and grammatical errors. As the classes show, it is difficult for students to complete such tasks without a Google translator. During the classes, students can be offered to edit the translation texts, correct grammatical, lexical and stylistic errors, creating their version in Ukrainian. It should also be noted that the use of translation programs when translating original specialized texts on architecture, which helps to speed up the translation process, immediately leads to quick forgetting. Therefore, it is recommended to do the translation manually. When translating, students become aware of the similarities and differences between foreign and native languages. Translation exercises students of architectural specialities of the Agrarian University learn to understand the difference in the grammatical structure of a foreign language and Ukrainian, and learn the skill of transferring the same thought in a speciality using different grammatical constructions. Frequent areas of difficulty in the translation of grammatical constructions include the translation of indefinitely personal and impersonal sentences, participles and participles, and the transfer of temporal relations between sentences. To avoid mistakes, a thorough study of the creation of written scientific texts in two languages is suggested. Here, the logical structure, style, and grammar are taken into account, special attention is paid to specialized vocabulary. Thus, students learn the skill of creating a full-fledged scientific article that will represent the results of their research in the future and meet the requirements for scientific papers around the world. Dictation, generally speaking, is one of the easiest forms of learning to write. This is reading the text aloud in such a way that the student can write down what is read to him word for word, after which the written text is checked and the errors are corrected. In the practice of teaching students of an architectural speciality, it is advisable to use specialized dictations, since there is a consolidation of active specialized vocabulary and characteristic grammatical structures through written dictation. Dictation is a useful and flexible activity in teaching a foreign language with an architectural orientation. You can use it when introducing a new structure, repeating and consolidating specialized vocabulary on architecture, and consolidating grammar. Dictation is an important and effective means of teaching writing in a foreign language, as well as a knowledge control tool. The third effective means of mastering a written foreign language of a specialized orientation is writing annotations and summarizing. Students of architectural specialities of agricultural universities learn to annotate and abstract specialized materials in a foreign language and master the appropriate terminology, as well as grammatical and lexical means characteristic of foreign documents and scientific correspondence.

Teaching writing remains a relevant and effective method despite the abundance of innovative teaching aids since it involves the visual, tactile and auditory, mental receptors of a person and contributes to the high-quality memorization of a specialized language.

ACKNOWLEDGMENT

This research received no grant from any organization. Personal thanks to Oleg Sergeevich Podgorodetskiy for the abundance of professional textbooks from his architecture library.

REFERENCES

Abramova, I. E., & Ananyina, A. V. (2010). Modernization of the language teaching system at the university. *Higher Education in Russia,* (8-9), 93-98.

Antonova, T. V. (2017). Dictation in teaching a foreign language as a means of forming an integrated skill. *Bulletin of the Kostroma State University. Series: Pedagogy. Psychology. Sociokinetics*, *23*(3), 164–169.

Bakhinskaya, I. A. (2021). Dictation as one of the ways for forming the discourse competence of technical students. *Linguistics and linguodidactics,* 14-17.

Baranova, N. A., & Ignatenko, E. V. (2016). Dictogloss as a means of activating the speech-cogitative activity of students in an English lesson. *Scientific and Methodological Electronic Journal "Concept", 12.* https://e-koncept.ru/2016/76137.htm

Baranova, N. A., & Ignatenko, E. V. (2016). Dictogloss as a means of activating the speech-cogitative activity of students in an English lesson. *Concept*, (S12), 1–5.

Bazanova, E. M. (2016). Scientific publication: write in English or translate? *Science Editor and Publisher, 1*(1-4), 17-24.

Belous, L. N. (2018). A modern approach to learning foreign words, or How to memorize new words in English. In Modern scientific research in the field of pedagogy and psychology (pp. 115-118). Academic Press.

Bezruchko, E. N. (2011). English for architects: Textbook for universities in the direction of "Architecture" (3rd ed.). Academic Press.

Blinov, M. A. (2017). *On the criteria of objectivity and subjectivity in assessing written communication skills in English of students of AF SPbGUP*. Academic Press.

Budagyan, I. G., Vishnevetskaya, N. A., & Romanova, O. N. (2015). Transfer of features of the translation of texts of architectural specialties. *Business. Education. Law*, (4), 351–354.

Chudley, R., & Greeno, R. (2006). *Building Construction Handbook*. Elsevier. doi:10.4324/9780080460864

Egorova, E. V., Krasheninnikova, E. I., Krasheninnikova, N. A., & Kuznetsova, O. Yu. (2018). Dictation in English (Total Dictation): Writing experience and results. Lingua Academica: Actual problems of linguistics and linguodidactics, 126-130.

Egoshina, N. G. (2022). Creating a cultural and educational space in the context of foreign language education. *Bulletin of the Perm National Research Polytechnic University. Problems of Linguistics and Pedagogy,* (3), 181-191.

Emmit, S., & Gorse, A. Ch. (2005). *Barry's introduction to construction of buildings*. Blackwell Publishing LTD.

Falaleeva, A. S. (2021). "Running dictation" as an effective method of conducting dictations in English lessons. *Modern Problems of Linguodidactics*, 126-129.

Frolova, N. A., & Aleshchanova, I. V. (2007). Training in abstracting and annotating professional literature in a foreign language in the complex training of a specialist. *Modern Problems of Science and Education*, (5), 11-11.

Golovina, N. V. (2012). The significance of the interdisciplinary aspect in the formation of translation activities among students of specialized classes of secondary school. *Bulletin of the Tambov University. Series. Humanities*, (9), 175–179.

Islamov, R. S. (2020). Artificial intelligence in information and communication technologies and its impact on foreign language teaching in higher education. *Philological Sciences. Questions of Theory and Practice*, *13*(12), 300–305.

Kalashnikova, A. A., & Akai, O. M. (2016). Methodology of translation and teaching translation from a foreign language in a non-linguistic university. *Modern High Technologies, 10*(2), 337-341.

Kokoreva, A. A. (2013). Corpus of parallel texts in foreign language teaching. *Bulletin of the Tambov University. Series. Humanities*, 2(118), 57–62.

Kolesnichenko, A. N. (2021). The use of digital technologies in teaching the vocabulary of a foreign language in higher education. *Pedagogy. Questions of Theory and Practice*, *6*(2), 298–303.

Konovalova, E. N., Marsheva, F. M., & Mubarakshina, F. D. (2014). *Professional English in use: Architecture. Part 1.* Kazan.

Kornienko, O. P. (2008). Terminological specificity of English texts of the sublanguage of architecture and construction. *Proceedings of the Russian State Pedagogical University*, 63-1.

Korotkikh, Ya. I., Mazalova, V. Yu., & Malinitskaya, E. P. (2019). Computer translation and its use in the educational process in a military university when studying a foreign language. Innovative educational space: Theory and practice of teaching foreign languages and Russian as a foreign language in higher education, 109-121.

Kuznetsov, V. A. (2006). Scientific adviser: Savinov AN. *Food Technology*, (2-3), 70.

Kuznetsova, N. G., Zaitseva, I. E., & Stepicheva, O. N. (2016). The Professional Language of Architects: A Systematic Approach to the Dictionary (Based on English, German and French). *Philological Sciences. Questions of Theory and Practice, 6*(60), 98-107.

Lebedeva, N. (2019). Information and communication technologies for teaching a foreign language regarding the modernization of public administration. *The latest information and communication technologies in the modernization of public administration: foreign and domestic experience: materials of II scientific practice seminar.* Dnipropetrovsk DRIDU NADU.

Lebedeva N. A. (2017). Information aspects in the context of innovative means of teaching a foreign language. *All-Ukrainian science and practice internet conference "Modern management: problems and development prospects on May 19, 2017"*, 235-238.

Lebedeva, N. A. (2018). Teaching English in the field of knowledge "Public management and administration" in an agrarian higher educational institution. In *European dimension of reforming public administration in Ukraine: materials of the International Scientific and Practical Conference.* Kyiv Publishing House "Personal".

Levina, L. V., & Levin, A. I. (2019). Ensuring continuity in the process of teaching the translation of authentic texts from a foreign language into Russian in a non-linguistic university. In *Foreign Language Teaching* (pp. 224–233). Modernity and Perspectives.

Likhacheva, O. N., & Bedenko, D. E. (2020). Teaching a foreign language in modern conditions in the context of a technical university. *Electronic network polythematic journal "Scientific works of KubGTU,* (5), 70-76.

Lipatova, V. V., & Litvinov, A. V. (2011). The concepts of equivalence and adequacy in teaching translation in higher education at the present stage. *Russian Journal of Linguistics*, (4), 105–113.

Melentieva, O. Y. (2009). The development of spelling vigilance at the initial stage of learning English. *Municipality: Innovation and experiment,* (2), 72-75.

Molodykh, O. V. (2021). Teaching spelling as a means of mastering written speech in English lessons. *Journal of Scientific Conferences, 2*(1), 81-82.

Musorina, O. A., & Sorokina, S. G. (2019). *Organization of educational work at the university at the level of the department: concept, content, methods for evaluating effectiveness. In Education: experience and development prospects. Chuvash Republican Institute of Education.* Publishing House LLC Publishing House Sreda.

Mustafina, F. Sh. (2014). Dictation as a means of improving graphic and spelling skills in teaching foreign languages. *English language. Everything for the teacher!, 32*(8), 6-7.

Nagiyatova, L. B. (2018). Learning to write: Dictation in an English lesson in a non-linguistic university. *Alley of Science, 4*(11), 185–188.

Osipova, V. S., & Putistina, O. V. (2021). The use of dictations in English lessons as a method of activation of students' speech thinking activity. *Collection of the Regional Scientific-Practical Conference-Student Scientific School of the MASU branch in Apatity*, 97-102.

Pahomkina, M. E. (2013). The role of translation in teaching a foreign language in the magistracy. *Proceedings of the Southern Federal University. Engineering Sciences*, (10), 99-103.

Petrosyan, I. V. (2019). The work program of the discipline "Practical course of the first foreign language" (for 1st year undergraduate students). *Collection of Programs,* 366-410.

Pilyugina, O. P. (2005). Scientifi c processing of professional texts//Technologies of teaching foreign languages in non-linguistic Universities. In *Collection of scientifi c articles of interuniversity scientifi c and technical conference.* Ulyanovsk: USTU.

Proskuryakova, N. A. (2013). Means of developing the abilities of students-future architects. *News of the Samara Scientific Center of the Russian Academy of Sciences, 15*(2-4), 914–916.

Romanova, O. N., Astafurova, T. N., Vishnevetskaya, N. A., & Surkova, E. V. (2022). Innovative technologies in teaching a foreign language to architect technologies in teaching a foreign language to architect students. *News of the Volgograd State Pedagogical University, 5*(168), 56–60.

Shalnev, N. A. (2014). Conducting dictations in foreign languages as a form of organization of classroom and extracurricular activities of students. *Languages of the peoples of the world and the RF,* 186.

Shcherbakova, I. V., & Timashova, M. V. (2019). The role of professionally oriented education in the development of dialogic and monologue speech in a foreign language in a non-linguistic university. *Baltic Humanitarian Journal*, *8*(3), 205–208.

Shevchenko, M. A., & Mitchell, P. D. (2013). Training of military translators in a civilian university (experience of the National Research Tomsk State University). *Language and Culture*, 1(21), 125–131.

Shishkina, T. G. (2020). The work program of the discipline "Practical course of the 1st foreign language (English) level "Advanced". *Collection of scientific and educational works,* 182-198.

Shor, G. A., & Galiullina, A. R. (2021). The role of the English Language in the international cooperation of landscape architects. *World of science, culture, education, 3*(88), 414-416.

Suprunov, S. E. (2019). Studying foreign languages in a non-linguistic university. Actual problems of the general theory of language, translation, intercultural communication and methods of teaching foreign languages, 122-126.

Takanova, O. V. (2008). Vocational-oriented teaching of a foreign language in non-linguistic universities. *Agroengineering*, (6), 49–51.

Trubina, Z. I. (2016). Written translation as a means of teaching a foreign language in a secondary school. *Pedagogical education in Russia*, (3), 171-175.

Tsarevskaya, I. V., & Volodina, M. S. (2016). Dictation as a valuable means of teaching a foreign language. *Modern High Technologies, 8*(1), 153-157.

Uvarova, I. V., & Erina, T. F. (2015). Interdisciplinary connections in teaching foreign languages in higher education. *International Journal of Applied and Basic Research*, (3-4), 724–726.

Volkova, N. P., Danko, O. Y., & Omelchuk, K. I. (2020). Formation of students' independent activity skills in English lessons through the use of CSR technology. Learning independence of the individual is the basis of lifelong education, 49-54.

Zhakhanova, Z. R. (2015). Structural and Semantic Features of Multicomponent Terms in Modern Russian and English on the Material of the Sublanguage of Architecture and Construction. *Problems of History, Philology, Culture, 2*(48), 274-279.

Chapter 6
Features of Teaching a Foreign Language to Lecturers of Higher Education Institutions

Olga Anatolievna Pasko
https://orcid.org/0000-0001-8322-834X
National Open Institute St. Petersburg, Russia

EXECUTIVE SUMMARY

The features of teaching a foreign language to lecturers of the higher school of non-linguistic specialties are considered. Relevant competencies are proposed. The features of the scientific text as the main product of the work are considered. A system of tasks is proposed in the context of simultaneous teaching of a foreign language and deepening of professional scientific competencies. Examples are given, including the development of an educational and methodological complex of the discipline "Land Management" (Science and Land). The unique experience of using modern foreign language teaching technologies in North Korea is described.

INTRODUCTION

The development of modern science and education stimulates the acceleration of the exchange of information between their participants - researchers, teachers of higher educational institutions and students. It provides an opportunity to exchange and acquire knowledge, skills and abilities. Along with the exchange of goods, products, and cultural values, this contributes to the globalization of the world community and the strengthening of its interconnectedness (Globalization, 2023). People from different countries and nationalities have a need and opportunity for

DOI: 10.4018/978-1-6684-6222-5.ch006

direct intercultural communication. The current level of development of technologies and technical means can satisfy this social demand.

Proficiency in a written foreign language allows a specialist to get involved in the work of attracting additional foreign scientific resources to their research. In Russia, a situation has developed where representatives of the older generation know a foreign language, sometimes not enough for successful work in higher educational institutions. This causes the need for additional training (independently or as part of advanced training). Currently, the methodological issues of teaching lecturers a professional foreign language are practically not developed. The main interest of researchers is focused on students and schoolchildren.

Thus, the relevance of this study is determined by:

- the need for theoretical development of issues of teaching tutors of non-linguistic specialities a professional foreign language;
- search for effective means of teaching oral and written foreign language communication;
- poor elaboration of the issues of teaching university lecturers to a professional foreign language.

The object of this study is to the features of the process of teaching English to lecturers of non-linguistic specialities in terms of advanced training.

The purpose of the study is to improve the professional training of non-linguistic teachers in the context of continuous education for successful integration into the international scientific and educational community.

To achieve this goal, the following methods of scientific research were used:

- study and analysis of scientific literature on the problem;
- summarizing the experience of teaching foreign language communication to university teachers in the speciality “Earth Sciences”;
- organized monitoring of the practical implementation of foreign language teaching of teachers in terms of advanced training.

The reliability of the results and conclusions of the study is justified by the theoretical and methodological elaboration of the problem under consideration and the use of a combination of correct qualitative methods for analyzing the material.

In the modern world, a foreign language plays the role of a necessary means of communication between teachers and scientists from different countries. A specialist who has little command of it either deliberately puts himself on the periphery of the scientific area, or must make efforts to achieve an acceptable level (fluent reading without a dictionary). As a rule, mastering and/or improving the level of knowledge of a foreign language takes place in parallel with the development of scientific research, outside of working hours.

Therefore, a specialist, as a rule, is critical and attentive to the acquired knowledge, evaluating them primarily from an applied point of view. He understands that he urgently needs, first of all, a certain minimum to enter the circle of colleagues who communicate in another language; that mastery of the language is not the goal of his efforts, and that the ability to grasp the essence of a foreign text on the fly will save a lot of time and effort in the future for selecting the necessary information.

As a rule, during advanced training courses, students receive a lot of useful information about foreign-language resources of specialized literature, grants, research teams, etc. They write review abstracts in their speciality and learn to navigate the international scientific world. This directly helps in the preparation of one's publications and at the same time provides a strategy for the development of research.

The main rule for motivating a specialist to teach a professional foreign language is an accurate understanding by the teacher of the "necessary and sufficient minimum", which will allow the specialist to get involved in the work of attracting additional foreign scientific resources to his research as early as possible. "Delay at the start" when mastering a foreign language, for example, a classic thorough analysis of its syntactic and morphological rules, excessive vocabulary often leads to a loss of interest among specialists and may be the reason for the termination of training.

Therefore, the choice of material and methods of teaching a foreign language for specialists should take into account their expectations, and resources (time and effort for learning, involvement in the scientific community), be formed taking into account this balance, accelerate the development of their scientific research and expand the circle of contacts with foreign colleagues. On the other hand, it requires reciprocal steps on the part of the teacher to understand the requests of the specialist. We are not talking about their scientific research, for example, in geology, geography, etc., but the teacher, for his part, also needs a certain amount of knowledge in the scientific direction developed by the specialist.

The superposition of these interests, the addition of the capabilities of a teacher and a specialist, makes it possible to bring scientific research to a qualitatively new level.

The purpose of the study is to analyze the features of teaching a foreign language to specialists who have little or no command of it, but at the same time are actively involved in scientific and educational activities. The level of foreign language proficiency among the teaching staff of Russian universities author of this chapter considers in such way: among the holders of a scientific degree in Russia, only 14% are fluent in English, and the rest (86%) find it difficult to use it in their professional activities, of which about a quarter easily overcomes them, a quarter experiences noticeable difficulties, 20% reads professional texts with a dictionary. 16% of graduated professors and teachers of Russian universities do not speak English at all. Teachers of agricultural sciences speak English worse, and teachers of natural sciences, mathematics, and medical sciences speak better. About 29% of degree holders speak German, and 12% speak French.

At the same time, the existing knowledge for successful international scientific and educational cooperation is not enough. As a rule, they were obtained without taking into account the professional orientation of their careers, so more than half of the degree teachers experience obvious difficulties in professional communication in a foreign language. For the full integration of Russian scientists into the scientific community, thoughtful systematic efforts are needed, in particular, the use of advanced training courses for teaching staff. In the author's opinion, the peculiarities of motivation for increasing the level of foreign language proficiency among the teaching staff of higher educational institutions in Russia may be considered in such way: among the modern means of teaching a foreign language to teachers and researchers of universities (hereinafter referred to as teachers), project-oriented learning has proven itself well. It provides for the possibility of choosing from a wide range of educational technologies that best correspond to the specifics of the scientific direction of teachers and at the same time are available to students.

An important point is the possibility of enhancing the effectiveness of their educational activities to help students in their training in professional activities, career planning, self-organization in learning, and teamwork (Khalilova, 2007).

The use of English helps to teach lecturers to "overtake without catching up", and to gain an advantage over colleagues in the conditions of an effective contract by mastering additional effective educational technologies. For example, knowledge of the latest achievements, and technologies for enhancing learning and developing creative activity help to prepare more successful students and graduates, and motivate them to participate in scientific competitions and conferences, thereby bringing up additional points to their leader in fulfilling the plan.

The use of relevant foreign literary sources allows one to prepare more significant scientific articles, publish them in higher-rated journals and, as a rule, receive certain bonuses (prizes, high scores for publication activity, etc.).

Foreign experience and its discussion in a foreign language make it possible to develop original educational materials for the disciplines taught based on the best experience of the world's leading universities in the original from primary sources.

As a successful example in this regard, there is the organization and its conduct of advanced training in professional English for university teachers developed at the National Research Tomsk Polytechnic University. It includes such educational technologies as active learning, project-oriented learning, non-formal learning (mentoring) in the course of solving real pedagogical and technical problems, independent work, mastering social skills and abilities, vocational guidance, counselling, etc.

On the part of teachers, the "life skills" technology is of great interest. During the implementation of them, there is a discussion of the latest scientific problems, an acquaintance from primary sources with the views of foreign colleagues, information resources, databases, etc. Teachers receive real tools to enhance the effectiveness of their scientific activities, teaching students self-organization, team and project work, career planning, etc. (Polat, Bukharkina, 2007).

The planned results of mastering the English language training program by scientific and pedagogical workers, for example, at the National Research Tomsk Polytechnic University are the following professional competencies, namely, abilities:

- to organize educational, research, project and other activities of students following the planned learning outcomes;
- to design, organize and implement the learning/self-learning process;
- to develop educational and methodological complexes of educational programs;
- to conduct an expert assessment of the results of the scientific and educational process and its organization.

The measured learning outcomes are abstracts, dictionaries, information resources, lecture notes, articles, presentations and other educational materials developed for the academic discipline of the professional training cycle, implemented in Russian and/or English.

In the course of teacher training at Tomsk Polytechnic University, there is a wide range of educational methods and teaching technologies are used, including competence-based and practice-oriented training (thinking development, information retrieval, analytical and creative activities, the principle of advanced independent work, case-study methods, project-based -oriented learning, debate method, etc.).

BACKGROUND

Knowledge of a foreign language contributes to faster and more successful integration into the foreign scientific environment and speedy adaptation to it (Khalilova, 2007). Therefore, a modern teacher should be prepared for international interaction both at the linguistic and professional levels (Senko, Frolovskaya, 2007). On the one hand, he receives a new store of knowledge and experience in the field of professional activity, on the other hand, he establishes contacts, has an exchange experience with foreign colleagues, prepares foreign publications and, as a result, advances in the ranking of university staff.

At the same time, the main goal of international communication in the learning process is to prepare a professional to use the acquired knowledge and skills aimed at developing the economy and production of the country's region (Khalilova, 2007).

The level of foreign language proficiency among the teaching staff of Russian universities is currently insufficient. The level of foreign language proficiency among the teaching staff of Russian universities is currently insufficient. Among the holders of a scientific degree in Russia, only 14% are fluent in English, the rest (86%) find it difficult to use it in their professional activities, of which about a quarter easily overcomes them, a quarter experience noticeable difficulties, 20% reads professional texts with a dictionary. 16% of graduated professors and teachers of Russian universities do not speak English at all (Shmatko, Volkova, 2023). Teachers of agricultural sciences speak English worse, and teachers of natural sciences, mathematics, and medical sciences speak better (Figure 1).

Figure 1. Knowledge of a foreign language by candidates and doctors of sciences (according to the field of science with which the main work is most related, %) in Russia
Source: Shmatko and Volkova (2023)

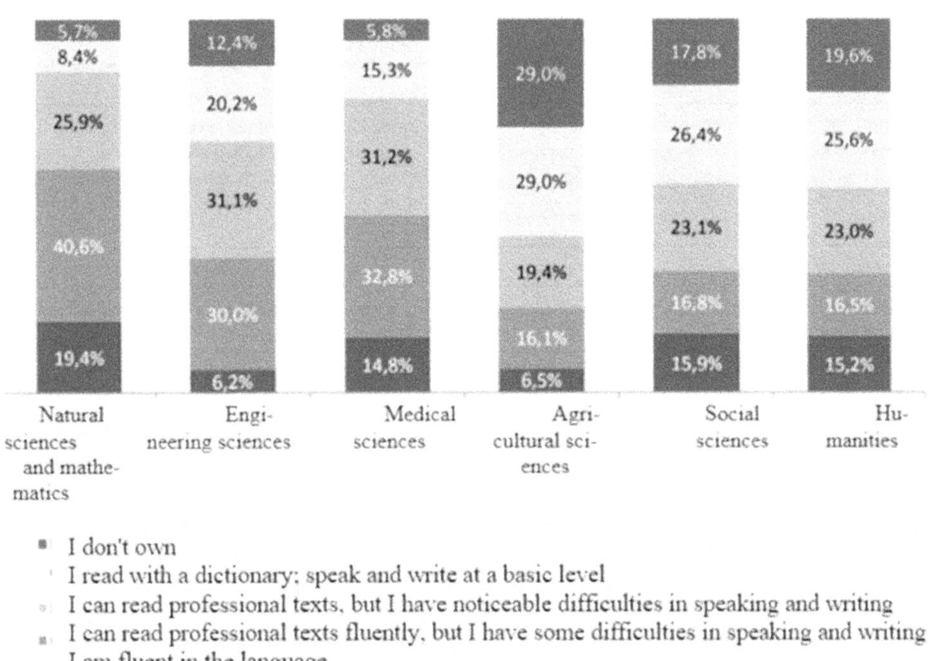

About 29% of degree holders speak German, and 12% speak French. At the same time, the existing knowledge for successful international scientific and educational cooperation is not enough. As a rule, they were obtained without taking into account the professional orientation of their careers, so more than half of the degree teachers experience obvious difficulties in professional communication in a foreign language. For the full integration of Russian scientists into the scientific community, thoughtful systematic efforts are needed, in particular, the use of advanced training courses for teaching staff (Shmatko, Volkova, 2023).

With the high demand for professional knowledge of a foreign language by university teachers, the corresponding learning technologies have been developed only for teaching students and teachers. A specially created structure of the educational process and a purposefully organized space for the formation of professional competencies contribute to the development of their need to master certain knowledge, skills and abilities. At the same time, the training of teachers in a foreign language has clear specifics, which should be taken into account in the process of their training.

Table 1. Features of teaching a foreign language to students and teachers

Student	**Teacher**
Studying Time	
As part of the compulsory educational process	In free time from the main job
Motivation	
Mastering the compulsory training program	Obtaining new information in the field of science and education; establishing direct contacts with foreign colleagues; preparation of foreign publications and grants; execution of an effective contract.
Psychological features	
Attitude towards discipline as a potential communication tool	Attitude towards discipline as a highly demanded communication tool
Use of technical means and technologies	
More confident use of innovative technologies: computers and telecommunications, smartphones, etc.	Combination of modern technical means and traditional paper media.
Use of electronic means of communication (voice and video communication, social services, e-mail, skype, software and hardware data processing systems).	
Wide	Limited

Peculiarities of motivation for increasing the level of foreign language proficiency among the teaching staff of higher educational institutions in Russia. Among the modern means of teaching a foreign language to lecturers and researchers of universities (hereinafter referred to as teachers), project-oriented learning has proven itself well. It provides for the possibility of choosing from a wide range of educational technologies that best correspond to the specifics of the scientific direction of teachers and at the same time are accessible to them.

An important point is the possibility of enhancing the effectiveness of one's educational activities to help students already learn their profession, plan their career growth, self-organization in learning, the attraction of new information and social resources, for teamwork.

For example, the use of a foreign language, first of all, in writing, helps teach the lecturers to "overtake without catching up", to gain an advantage over colleagues under the conditions of an effective contract by mastering additional effective educational technologies. A certain pool of scientific publications is an admission to submitting a research grant, forming an appropriate scientific team, and directing and coordinating its activities (Table 2).

Knowledge of the latest achievements and technologies for enhancing learning and developing creative activity helps to prepare more successful students and graduates and to motivate them to participate in scientific competitions and conferences. As a rule, this brings their leader additional points in the implementation of the plan and is financially encouraged by the leadership of the university. The use of relevant foreign literary sources allows one to write more significant scientific works and its discussion in a foreign language makes it possible to develop original educational materials in the lecturers' programs for their students.

Table 2. The role of a written foreign language in the development of scientific and educational competencies of a teacher of a higher educational institution

Formation of a Scientific School	
Obtaining additional funding. Overfulfillment of an effective contract.	Participation and victories of students in conferences and olympiads. Postgraduate training.
Fulfillment of filing requirements grant applications. Getting a grant.	Preparing More Successful Students
Increasing the citation index	Involvement of up-to-date international information in the educational process
Publication over valuable scientific articles	Use of the latest scientific information in educational activities.
Improving written foreign language proficiency	

Text Analysis in the System of Teaching a Foreign Language to Teachers of Non-Linguistic Disciplines

One of the means of optimizing the process of teaching a foreign language, the formation of professional and communicative competence of students is the use of textbooks and teaching aids created based on such principles as a text-centric approach, a semantic approach, multi-level text analysis, abstracting and annotation strategies, a comparative analysis of two and more texts, the use of psychological and semantic translation mechanisms, writing analytical essays and others (Baranova, Khalilova et al., 2017).

To understand the meaning encoded in texts in a foreign language, it is necessary to use all possible methods and mechanisms for text analysis, translation, abstracting and annotation.

According to some authors (Khalilova, 2007; Van Dijk, 1989), the effective impact of the text on the recipient is due to many factors: the logical and semantic organization of the text, its thematic division, compositional structure, ideological and thematic structure, stylistic devices and expressive means; reasons that prompted the author to use just such means.

Currently, the text is considered a unit of learning, an integral information space. A grammatically and lexically correct phrase that is not correlated with the norms of speech behaviour of native speakers can disrupt communication (Van Dijk, 1989). The task is to understand the external and internal structure of the text, which underlie the teaching of semantic folding, probabilistic forecasting and other psycholinguistic mechanisms that are indispensable in teaching abstracting and translation, as well as in carrying out any research work.

At present, there are four types of reading (Khalilova, 2007).

New pedagogical approaches are put forward and actively applied - competence-communicative, personal-activity, phenomenological and others, which develop, however, along with the traditional approach, which has confirmed its effectiveness over the past decades (Federal State Educational Standard ... 2018).

By analogy with other federal state standards, the following competencies can be offered for university teachers who improve their knowledge of a foreign language:

- English language proficiency at an advanced level by improving their competencies, skills and abilities in the field of text analysis, summarizing and annotating, as well as translation;
- formation and development of competencies, which are based on the ability to communicate in writing in a foreign language to solve the problems of scientific research;
- ability to use foreign languages to solve professional issues;
- the ability to clearly, logically correctly and reasonably build written speech in the professional field;
- the ability to perform written and oral translations of professional materials from a foreign language into Russian and from Russian into a foreign language;
- possession of techniques for establishing professional contacts and developing professional communication, including in a foreign language;
- the ability to master the skills of public speaking both in front of Russian and foreign audiences;
- the ability to master the basic skills of reading scientific texts in a foreign language;
- possession of standard methods of computer typing in a foreign language;

Working with text involves the ability to respectfully and carefully treat the historical heritage, cultural traditions and scientific achievements of foreign peoples, understanding social and cultural differences; possession of a culture of speech, the basics of scientific and diplomatic etiquette; the ability to adapt to work in international groups, the ability to work with materials from various foreign information resources; the ability to compile literature reviews on given topics, to find, collect and initially generalize factual material, making reasonable conclusions; the ability to give a comprehensive description of a scientific problem (Senko, Frolovskaya et al., 2007).

RESULTS AND DISCUSSION

To develop all the competencies, work is required in the following areas:

In the field of expert and analytical activities:

- Development of skills in analytical reading and translation of original scientific foreign-language texts for solving professional problems;
- Development of the ability to understand the essence and meaning of the information received.

In the field of educational activities:

- Development of translation and interpreting skills;
- Development of a communicative-pragmatic approach to the analysis of a foreign-language scientific text;
- Identification and analysis of lexical-phraseological and structural-compositional features of foreign-language scientific texts;
- Preparation, reviewing and editing of scientific and popular science works (based on knowledge of a foreign language) and advising the authors of works.

In the field of research activities:

- Development of independent research skills;
- Development of the ability to use knowledge of the principles of compiling scientific and analytical reports, reviews, information notes and explanatory notes;
- Developing the ability to understand, present and critically analyze information.

They correspond to different types of presentation of information (Table 3).

Understanding foreign language text (decoding information).

Viewer assignment of the text to a certain speech form: text description; reasoning text; comment text; understanding of its scientific value.

Table 3. Correspondence to the types of presentation of information to types of reading

Type of Reading	Type of Presentation of Information				
	Annotation	Abstract	Written Translation	Interpretation	Summary
Viewer	+				
Introductory	+	+		+	+
Search engine	+				
Studying		+	+		

Introductory reading:

1) Definition of the main problem of the text and its target setting
2) Extracting text information based on the "golden core" of its content search reading
3) Search for facts, formulas, conclusions, bibliography

Studying

4) Analysis of the full information of the text, and selection of the most valuable materials.

There are the following functions of the text in a foreign language: information, transformational, systematizing (Baranova, Khalilova et al., 2017), integrating, coordinating controlling, motivating, communicative, and educational.

Taking into account the presented features of the scientific text, the author of this chapter has proposed the following tasks for teaching a foreign language to lecturers of non-linguistic specialities.

TASK 1. WRITE AN ESSAY ON THE TOPIC OF THE SCIENTIFIC RESEARCH OF THE TEACHER

An abstract is a text containing an accurate and brief summary of the content of one or more scientific publications without comments. The structure of the abstract corresponds to the structure of the publication. As a rule, it includes relevance, purpose, object, subject, research methods, main results, discussion, conclusions and a list of references.

The functions of the abstract are informational and search.

Methods for processing scientific information of the abstract

- Extraction,
- Paraphrasing,
- Interpretation.

The recommended length of the text of the abstract is 850 printed characters (determined by the requirements of the journal).

1. Translate keywords into a foreign language.
2. Find relevant journals and monographs in Russian and foreign languages in information databases.
3. Familiarize yourself with their abstracts and source data.
4. Carry out a primary selection of material, giving preference to publications that best meet the scientific interests of the author, as well as published in the last 5 years.
5. Save the original data of the publication for quick retrieval.
6. Copy interesting illustrations, tables, formulas, and conclusions.
7. Systematize the material by headings.
8. Write separate sections of the abstract.
9. Collect an abstract, structure it and arrange it.

TASK 2. WRITE AN ACADEMIC SUMMARY (CURRICULUM VITAE) ABOUT YOURSELF

An academic resume (CV, Curriculum Vitae) is a questionnaire that contains information about education, skills and academic achievements. It is usually required by institutes, universities and some employers. CV characterizes the biography, qualifications and achievements of the teacher.

The academic summary consists of the following sections:

- Education (university, faculty, speciality, years of study, city, country - in reverse chronological order);
- Work experience (organizations/institutions; positions, years of work). The information is grouped in reverse chronological order (last position at the top). It is recommended to focus on personal achievements;
- List of scientific publications by sections (scientific articles, monographs, conference materials, educational and teaching aids, inventions);

- List of conferences and seminars in which the author spoke (speaker, lecturer, etc.);
- Grants and scholarships in which the author took part (supervisor, performer);
- Language certificates;
- Membership in organizations / scientific communities;
- Extra education;
- Other sections:

Skills: this is a simple and clear way to give an idea about the author;
Volunteer experience:
Links: at the request of the employer.

- Personal qualities.

Good resume writing examples are here (Curriculum Vitae) – Resume Templates, 2023; Curriculum Vitae (CV) Samples, 2023; How to write a CV. 2023; CV and Cover Letter Advice, 2023).

TASK 3. CREATE YOUR SCIENTIFIC DICTIONARY

Learning words from your dictionaries is very effective since the teacher enters into the dictionary only those special words and terms that he does not know, but wants or should know for one reason or another. When new words are added to the dictionary, they are involuntarily memorized. It is recommended to have several of your dictionaries on various scientific topics and maintain them, that is, add new words and at the same time repeat the old ones.

It is recommended to create your dictionary in EasyWords or

MS Excel. The columns should be in the following order: word, transcription, translation, and note.

Execution example (table 4).

Table 4. Dictionary

N°	Word	Transcription	Translation	Note
1	Современная	mɒdn	Modern	
2	Динамическая	daɪˈnæm.ɪk	Dynamic	
3	Историческая	hɪˈstɒrɪkl	Historical	
4	Геология полезных ископаемых	ˈmɪnərəl	Mineral geology	
5	Региональная	ˈriːdʒən	Regional	

What is transcription? Phonetic transcription is a graphic recording of the sound of a word; pursues the goal of accurate graphic recording of pronunciation. Each sound must be separately recorded in the recording. Phonetic transcription is written in square brackets; special phonetic symbols are used for recording.

When completing the task, you should use the resources (Google Translate, 2023; Transcription, 2023) for pronunciation and translation of English words online

It gives transcription, pronunciation and translation of English words online.

To use it, you need to enter a word in the service window and click "Search". After a short pause, he gives out a transcription of the English word, pronunciation and translation. For convenience, two options are offered: British and American. You can also listen to pronunciation options online.

TASK 4. LIST OF DISCIPLINES

Compile a list of disciplines taught by the teacher according to the levels of student learning and place them in direct chronological order

Example.

Educational-methodical work

Lecture courses developed by Maria A. Ivanova

2010-2013 N. State University Lectures and practical activities in courses: "Physiology of plants and basic biochemistry", "Microclonal propagation of plants"

2013-2019 N. Agricultural Institute Lectures and practical activities in courses: "Selection and seed industry", "Botany", "Ecology", "Vegetable Growing", "Fruit Growing", and "Landscape Gardening".

2019 National Research N. Polytechnic University. Lecture courses: " Basic land management", "Land parcel management", "Cartography", " Basic natural resource management", "Agricultural production economics, management and Technology", "Student's research activity", "Regional land management", "Land management design (Interfarm land management design. Intrafarm Design. Land parcel use planning and land resource management), "Earth resources", "Land resources management", "Teaching practicum", "Cadastral Register", "Work experience internship"," Modern issues of land management, Cadastral Register and land monitoring", "Research practicum", "Research work of a post-graduate student", "Geo-informational and land-informational systems".

Courses delivered in NPU (lectures, practical classes and laboratory work developed by Ivanova)

Bachelor's degree program

Basic natural resource management

1. Land parcel management
2. Cartography
3. Basic land management
4. Agricultural production economics, management and technology
5. Student's research activity
6. Regional land management
7. Land management design (Interfarm land management design. Intrafarm design. Land parcel use planning and land resource management).
8. Land resources

Master's degree program

1. Land resource management
2. Teaching practicum
3. Student's research activity
4. Cadastral Register
5. Work experience internship
6. Modern issues of land management, Cadastral Register and land monitoring

Post-graduate program

1. Work experience internship
2. Teaching practicum
3. Research practicum
4. Research work of a post-graduate student
5. Geo-informational and land-informational systems

Coordinator of Basic Education Programs

Coordinator of basic education programs Master's degree programs. Speciality 21.04.02 "Land management and Cadastres" Course "Land resource management". Qualification: Master of land management. Postgraduate program. Speciality 05.06.01 "Earth science", Course 25.00.26 "Land management, Cadastre and Land monitoring". Qualification: Researcher. Teacher.

TASK 5. COMPOSE THE LIST OF PUBLICATIONS

Compile a list of the author's publications based on the requirements of a particular publication.

Example

1. Pasko O.A., Kovyazin V.F., Zakharchenko A.V., Lebedeva N.A. Spatial organization of forest fund// IOP Conference Series: Earth and Environmental Science, 2020, 574(1), 012061.
2. Tokareva O.S., Alshaibi A.D.A., Pasko O.A. Assessment of restoration dynamics of burnt forest area vegetation using Landsat satellite data // Bulletin of the Tomsk Polytechnic University, Geo Assets Engineering, 2021, 332(7), pp. 191–199.
3. Pasko O., Staurskaya N., Zakharchenko A., Zharnikov V., Larionov Y. Bio-Farming as the Basis of Environmentally-Sustainable Arable Farming at the Time of Global Warming //Research Anthology on Strategies for Achieving Agricultural Sustainability, 2022, pp. 600–627.
4. Pasko O.A., Zakharchenko A.V., Kovyazin V.F. Analysis of land management of the forest fund // Bulletin of the Tomsk Polytechnic University, Geo Assets Engineering, 2021, 332(2), pp. 127–138.
5. Lebedeva, N. A. (2022). Teaching of Foreign Language in the Educational Process of Public Administration Personnel in Ukraine. In O. Kulaç, C. Babaoğlu, & E. Akman (Eds.), *Public Affairs Education and Training in the 21st Century* (pp. 294-310). IGI Global. http://doi:10.4018/978-1-7998-8243-5.ch019

TASK 6. SYLLABUS AND TEACHING AIDS DEVELOPMENT FOR THE COURSE "PROFESSIONAL TRAINING IN ENGLISH"

Compile a Calendar Plan and Assessment Measures for the Course;

Exam Paper;

The working program of the discipline.

Example

Annotation

This work is devoted to the development of elements of educational and methodological support of the discipline "Professional training in English" for students of the Engineering School of natural resources Department of Geology, studying in the direction of 05.06.01 Earth Science profile 25.00.26 "Land Management, cadastre and land monitoring" (Level of higher education training of scientific and pedagogical personnel in graduate school).

The program is designed for a three-year course for students of recruitment in 2021. The discipline is taught in the second year (3-4 semesters). The paper presents the working program of the discipline, examples of evaluation tools, types of training activities and selected forms of organization of the process.

Table 5. Syllabus for the course "Land Management, Cadastre and Monitoring of Land" course overview

Course Objectives	The course is aimed at training specialists able to: (a) to value land resources (b) to apply regulations to land administration; be able to give examples of compliance and violations; (c) to predict the scenario of development of land condition through the use of data monitoring; (d) to evaluate a variety of projects and select the best ones for a given situation; (e) to demonstrate an understanding of the main trends in land management, cadastre and monitoring in Russia and the world; (f) to analyse the state of natural resources using modern technology and techniques; (g) to model the state of the earth's surface and the main processes occurring on it (mathematical, functional and technological models); (h) to ask and answer problematic questions in a professional manner. The course goals are in compliance with the goals of "Land management, cadastre and monitoring of land" postgraduate program within 25.00.26 land Management, cadastre and monitoring.
Learning Outcomes Результаты обучения	Upon the course completion, postgraduate will be able to: to acquire the knowledge necessary for the development of scientific approaches to land management, as well as the organization of effective use of the land Fund and monitoring its status; to learn the skills of obtaining, collecting and processing information for the effective solution of practical problems of the state real estate cadastre, land management and land monitoring.
Course Outline	This program is based on the continuity of programs in the higher education system and is intended for graduate students of TPU, trained under the master's program, who have attended the relevant courses and have positive assessments on them. It is based on the provisions reflected in the curricula of these levels. Discipline "Land Management, cadastre and monitoring of land" is necessary in the preparation of scientific and qualification work of a graduate student and preparation for the candidate's exam.
Prerequisites (if avaliable)	For the development of the discipline "land Management, cadastre and land monitoring" requires knowledge and skills acquired by students as a result of the development of a number of previous disciplines, such as: Information computer technology, Modern geosystems and technologies, Monitoring of land and natural resources, Planning, forecasting of land use, Research work in the semester, Pedagogical practice, Fundamentals of pedagogical activity, Remote research methods in land management, Management of innovations
Course Structure	The course "Land Management, Cadastre and Land Monitoring" includes 5 Modules comparing practices. ***Module 1. Improvement of land use in Russia on modern ethane*** Practices 1. Characteristics of the current state and use of the land Fund. Practices 2. Problems of rational use of the land Fund in modern conditions and ways to solve them Practices 3. Land legal relations as a basis for the formation of land management systems, cadastre and land monitoring ***Module 2. Cadastre*** Practices 4. Modern problems of real estate cadastre and their solutions. Practices 5. Theoretical principles, goals, objectives and principles of the GKN. Practices 6. Principles of collection, documentation, accumulation, processing and storage of information about land plots. Practices 7. Information support of GKN ***Module 3. Land monitoring*** Practices 8. Development of state land monitoring. Modern technologies of land monitoring. Practices 9. Development of state land monitoring. ***Module 4. Land use development strategy in Russia*** Practices 10. Optimization of forms of economic use of land. Practices 11. Protection of lands. Practices 12. Planning and organization of rational use and protection of land. ***Module 5. Organizational, legal and economic basis for the use of land resources*** Practices 13. Scientific basis of land use organization Practices 14. Legal mechanism of formation of the state land cadastre system Practices 15. Economic regulators of rational land use. Practices 16. State regulation of land relations in market conditions. Practices 17. Legal support of land management and land cadastre actions Practices 18. Interaction of the state real estate cadastre, state monitoring of land and land management. Practices 19. Prospects for the development of land management, cadastre and land monitoring

continued on the following page

Table 5. Continued

Facilities and Equipment	For practical training: Multimedia projector – 1 PC., computer – 1 PC., speakers – 1 PC., screen – 1 PC (Tomsk, Lenin Ave., 2/5, educational and laboratory building N° 20, room 120). Multimedia projector-1 PCs., speaker – 1 PCs., - screen – 1 PCs computers – 13 PCs (Tomsk, Lenina, 2/5, educational laboratory building N° 20, auditorium 101-a)
Grading Policy	Evaluation of the quality of learning performance and achievements for the discipline (module) during formative and summative assessment is performed in compliance with "TPU Regulations for Formative and Summative Assessment" (rector`s order # 59/од dated as 25.07.2018). The maximum mark for one discipline for one semester is 100 points including: mastering the discipline content (exam type – pass/fail exam; graded credit test); evaluated in formative assessment and summative assessment (interim attestation) is 100 points; mastering the discipline content (exam type – exam); evaluated in formative assessment is 80 points and summative assessment (interim attestation) is 20 points; The maximum mark for the course final project (course final paper) for one semester (depending on what is applicable in compliance with curriculum) is 100 points including: – marks for formative assessment is 40 points, marks for summative assessment (interim attestation, defense) is 60 points. Assessment measures for formative assessment designed for discipline (module) corresponding to learning performance and achievements are stated in the Attachment "Calendar Plan and Assessment Measures for the Course".
Course Policy	To be allowed to take the final test, a student must attend all classroom activities and submit reports for all laboratory and practical sessions or other associated tasks applicable to the course.
Teaching Aids and Recourses	Compulsory Readings: [1] Use Harvard Referencing Style. https://curve.coventry.ac.uk/open/file/bdfb947c-9d43-48d3-8ec8-f511682e1dd1/1/The%20CU%20Guide%20to%20Referencing%20in%20Harvard%20Style.pdf [2] Google translate (2023). https://translate.google.com/?hl=ru Further Readings: 1. Bespalenko, O. N. Modern problems of land management, cadastre and land monitoring: textbook/ O. N. Bespalenko. Voronezh: VGLTA, 2012. – 140 p. 2. Walger J. W.. Real estate as an object of capital investment / J. V. Walger; nauch. hands'. I. G. Vidyaev // Problems of market economy management interregional collection of scientific works: / national research Tomsk Polytechnic University (TPU) ; ed. I. E. Nikulina ; L. R. Tukhvatulina ; N. V. Cherepanova . – 2015 . – Vol. 2 . – P. 185-188. 3. State real estate cadastre: textbook / A. A. Varlamov, S. A. Galchenko; agricultural education. – Moscow: Koloss, 2012. – 679 p. 4. Slezko V. V. land Management and land use management: textbook for universities / V. Slezko, E. V. Slezko, L. V. Slezko. – Moscow: Infra-M, 2013. – 202 p. 5. Murzin, A. D.. Real estate: Economics, valuation and development: textbook for universities / A. D. Murzin. – Rostov-on-don: Phoenix, 2013. – 382 p. 6. Typology of real estate: textbook for universities / Ia Sinyansky [et al.]. – 2nd ed., erased. – Moscow: Academy, 2014. – 318 p.
Instructor(-s)	Olga A.Pasko email: oap@tpu.ru

Table 6. Samples of control materials and tests. Exam Paper No10. The Course Title: Graduate direction 05.06.01 "Earth science", Profile 25.00.26. "Land Management, Cadastre and Monitoring of Land"

No	Questions	Mark
1	Land is a natural resource and the main means of production in agriculture and forestry.	10
2	Land protection is the basis of an integrated approach to land use, taking into account their zonal and regional characteristics (prevention of land degradation, ecological state of land, protection from water and wind erosion, from contamination of agricultural land, reclamation of disturbed land).	10
3	Agroecological monitoring in intensive agriculture; components of agroecological monitoring; goals, objectives, structure, biogeochemicalapproaches to monitoring	10
4	Task. To systematize innovative developments in land management, cadastre and monitoring of land by object, means, purposes, etc. Give examples.	10
	Total marks	*40*

Table 7. Glossary

N°	English	Русский вариант	Значение слова
1	**Land plot rent**	**Аренда земельного участка**	Форма договорного, срочного и возмездного пользования земельным участком.
2	**Tenants of land**	**Арендаторы земельных участков**	Лица, владеющие и пользующиеся земельными участками по договору аренды, договору субаренды.
3	**The balance of the land**	**Баланс земельных угодий -**	Система показателей, характеризующих наличие и изменение земельных угодий по категориям земель, категориям землепользователей и видам угодий.
4	**Bioindicator**	**Биоиндикатор**	Группа особей одного вида или сообщества, по наличию или по состоянию которых, а также по их поведению судят о естественных и антропогенных изменениях в среде.
...	**...**	**...**	...
97	**Environmentally friendly product**	**Экологически чистый продукт**	Продукт, который, по заверениям его производителей, не нанесет ущерба окружающей среде.
98	**Environmental standard**	**Экологический норматив**	Величина антропогенной нагрузки, рассчитанная на основании экологических регламентов и получившая правовой статус.
99	**Ecological safety**	**Экологическая безопасность**	Допустимый уровень негативного воздействия природных и антропогенных факторов экологической опасности на окружающую среду и человека.
100	**Ecological expertise**	**Экологическая экспертиза**	Установление соответствия документов и (или) документации, обосновывающих намечаемую в связи с реализацией объекта экологической экспертизы хозяйственную и иную деятельность.

Table 8. Evaluation methods Learning outcome: At the end of the module the learner will be expected to write clear, well-structured abstract to the scientific topic on land management, cadastre and land monitoring, with a clear structure and a high degree of grammatical accuracy based on the problem provided. Assessment method – the students will write an abstract to the scientific topic on land management, cadastre and land monitoring be chosen by them in the frame of their research work. Assessment rubric (1)

Abstract Criteria	**5/*Excellent***	**4/*Very Good***	**3/*Adequate***
Descriptive Title	Title is complete and informative, and written in the academic style.	Title is complete, but very long or contains words, that can be missed.	Title is informative and appropriate, yet is somewhat incomplete, contains errors, or is written in an unscientific style.
Introductory Statement	Clear, concise, engaging; Describes, connects the topic to literature and purpose of work. It is written in the academic style.	Clear, but not engaging; Attempts to connect to literature. It is written in the academic style.	Unclear. Attempts to connect to literature or is written in an unscientific style.
Purpose	Clear, concise, andrelevant.	Clear but not concise; Might contain irrelevant or unimportant information; lacks specifics.	Clear but contains irrelevant or unimportant information.
Methodological Approach	Connected to the purpose of the abstract; Identifies method used to support thesis or answer the research question.	Connected to the purpose of the abstract; Not clear identifies methods of experiments why they were used (advantages, disadvantages, etc.), how they solve the research question.	Unclear or not connected to purpose of abstract.
Results	Clear, connected to the purpose of the abstract; Provides explanation of what was expected, discovered, accomplished, collected, produced.	Result section is mostly complete, but some relevant results are omitted, or incorrect units or significant figures are used.	Attempts to present findings but might be unclear; or some information missing.
Summary of Conclusions	Conclusions are made based on the results, any accepted values are given for comparison, and percent error values are provided	Most Conclusions are presented.	Conclusion section is generally complete, but some relevant conclusions are omitted
Compliance with the rules of abstracts (each event has its own)	Writing appropriate for the scientific event. Using necessary technical terms. Defines all acronyms at first use. All requirements of the organizers are taken into account.	Writing appropriate for the scientific event, but need some corrections.	Not all requirements of the organizers not taken into account (references, figures, tables, etc.), but abstract can be corrected and can be send again.
Grammar	The abstract is free from serious grammatical or mechanical errors. Appropriate verb tense (present/past tense when talking about the study.	Few grammatical errors or typos; Mixed verb tense. The sentences are mostly clear but lack complexity, and demonstrate some variety. Style is generally appropriate to academic writing.	The abstract contains several errors in spelling, grammar, punctuation, and/or syntax that may somewhat distract from the reading of the abstract.

Assessment rubric (2)

Abstract Criteria	**2/*Weak***	**1/*Verypoor***
Descriptive Title	Title is neither informative nor appropriate	Missing
Introductory Statement	Unclear. Doesn't connect to literature. It is written in an unscientific style.	Missing

continued on the following page

Table 8. Continued

Abstract Criteria	5/*Excellent*	4/*Very Good*	3/*Adequate*
Purpose	Unclear; Contains irrelevant or unimportant information.		Missing
Methodological Approach	Not mentioned but implied; or, not appropriate for purpose of abstract.		Section is absent or not relevant to the experiment
Results	Unclear; Or not related to the purpose of the abstract; Or misinterpretation of results.		Missing
Summary of Conclusions	Less than half of Section complete, or not relevant to the experiment.		Section is absent, less than half complete, or not relevant to the experiment.
Compliance with the rules of abstracts (each event has its own)	Abstract is not in frame of session and contains characters more than necessary, More than half of requirements of the organizers not taken into account (references, figures, tables, etc.).		Abstract is written in unappropriated way.
Grammar	Many grammatical errors, typos but they do not impede understanding; inappropriate verb tense		Grammatical errors, typos impede understanding; inappropriate verb tense

Table 9. Calendar plan and assessment measures for the course

Land Management, Cadastre, and Monitoring of Land								
Students Group No A8-78			Graduate Student Training Program in Earth Science 05.06.01	Profile 25.00.26				
School of Natural Resources	Department of Geology	Semester 3 и 4	Autumn, 2018; Spring, 2019	Academic year 2		2018-2019		
Instructor Pasko O.A.								
Week of Studies	**Learning Outcomes**	**Topics and learning formats**	**Delivery hours**		**Assessment Measures**			
			Face-to-face	**Independent studies**	**Abstract**	**Presentation**	**Test**	**Points**
		Module 1. *Improving land use in Russia on modern ethane*						
1 2-3 4-5	To acquire the knowledge necessary for the development of scientific approaches to land management	**Seminar 1.** *Characteristics of the current state and use of land*	2	8	1			**3**
		Seminar 2. *Problems of rational use of land in modern conditions and their solutions*	2	10		1		**3**
		Seminar 3. *Land legal relations as a basis for the formation of land management systems, cadastre and land monitoring*	2	10			1	**3**
		Tutorials: [1], 1,2,3.						
		Module 2. *Cadastre*						

Table 9. Continued

Land Management, Cadastre, and Monitoring of Land									
6 7	To acquire the knowledge necessary for the effective use of the land Fund and monitoring its status	**Seminar 4.** *Modern problems of the real estate cadastre and ways to solve them.*	2	8		1		**3**
		Seminar 5. *Theoretical provisions, goals, objectives and principles of the State Real Estate Cadastre.*	2	10	1			**3**
8		**Conferenceweek 1**						**2**
9 10-11	To learn the skills of obtaining, collecting and processing information for the effective solution of practical problems of the state real estate cadastre, land management and land monitoring.	**Seminar 6.** *Principles of collection, documentation, accumulation, processing and storage of information about land plots.*	2	10			1	
		Seminar 7. *Information support of the State Committee.*	2	10		1		**3**
		Tutorials: [1], 1, 4. 5						
12- 15		**Module 3.** *Land Monitoring*			1			**3**
		Seminar 8. *Modern technologies of land monitoring.*	2	14			1	**3**
		Seminar 9. *Development of state land monitoring.*	2	14		1		**3**
		Tutorials: [1], 2, 3						
16		**Conferenceweek 2**						**2**
		Module 4. *Land use development strategy in Russia*						
17- 23	To master the methods of cadastral valuation of real estate	**Seminar 10.** *Optimization of forms of economic use of land.*	2	16	1			**3**
		Seminar 11. *Land protection*	2	16		1		**3**
		Seminar 12. *Planning and organization of rational use and protection of land.*	2	16			1	**3**
		Tutorials: [1], 2, 5						
24		**Conferenceweek 3**						
		Module 5. *Organizational, legal and economic bases of use land resources*	2					

Table 9. Continued

Land Management, Cadastre, and Monitoring of Land								
25 26 26-27 27-28 29-31	To get acquainted with the legal framework of land management and land monitoring. To learn their uses. To master methods of detection of violations and their elimination.	**Seminar13.** *Scientific bases of land use organization*	2	10			1	**3**
		Seminar14. *The legal mechanism for the formation of the state land cadastre system*	2	10		1		**3**
		Seminar 15. *Economic regulators of rational use of land.*	2	12	1			**3**
		Seminar16. *State regulation of land relations in market conditions.*	2	10		1		**3**
		Seminar 17. *Legal support of land management and land cadastral actions*	2	10			1	**3**
		Seminar 18. *The interaction of the state cadastre of real estate, state monitoring of land and land management.*	2	10	1			**3**
		Seminar 19. *Prospects for land management, cadastre and land monitoring*	2	10		1	1	**6**
		Tutorials: [1], 1 - 5						
		Total hours	**324**	**252**				
		Total for the formative assessment No 4			**60**			**max**
		Exam						**40**

Compulsory Readings:

[1] Use Harvard Referencing Style. Available at: https://curve.coventry.ac.uk/open/file/bdfb947c-9d43-48d3-8ec8-f511682e1dd1/1/The%20CU%20Guide%20to%20Referencing%20in%20Harvard%20Style.pdf

Further Readings:

1. Bespalenko, O. N. Modern problems of land management, cadastre and land monitoring: textbook/ O. N. Bespalenko. Voronezh: VGLTA, 2012. – 140 p.
2. Walger J. W.. Real estate as an object of capital investment / J. V. Walger; nauch. hands'. I. G. Vidyaev // Problems of market economy management interregional collection of scientific works: / national research Tomsk Polytechnic University (TPU) ; ed. I. E. Nikulina ; L. R. Tukhvatulina ; N. V. Cherepanova . – 2015 . – Vol. 2 . – P. 185-188.

3. State real estate cadastre: textbook / A. A. Varlamov, S. A. Galchenko; agricultural education. – Moscow: Koloss, 2012. – 679 p.
4. Slezko V. V. land Management and land use management: textbook for universities / V. Slezko, E. V. Slezko, L. V. Slezko. – Moscow: Infra-M, 2013. – 202 p.
5. Murzin, A. D.. Real estate: Economics, valuation and development: textbook for universities / A. D. Murzin. – Rostov-on-don: Phoenix, 2013. – 382 p.
6. Typology of real estate: textbook for universities / Ia Sinyansky [et al.]. – 2nd ed., erased. – Moscow: Academy, 2014. – 318 p.

Table 10. Assessment Descriptors

Formative Assessment, contribute to 100% to the final mark	Summative Assessment, mark		Mark Range	Quality of Performance	Grade	Assessment Descriptors
	Exam / credit	Course final paper and project				
90%÷100%	39-40	57÷60	96÷100	Excellent	A+	Excellent systemic knowledge of subject, key aspects of area of study, very high degree of skills and expertise in practical activities. Learning outcomes are achieved and evaluated withinthe maximum mark range for the course.
			90÷95		A	
70% - 89%	35-38	52÷56	80÷89	Good	B+	Very good understanding of subject, key aspects of area of study, high degree of skills and experience in practical activities. Learning outcomes are achieved andnone of these is evaluated withinthe minimum mark range for the course.
	31-34	46÷51	70÷79		B	
55% - 69%	22÷30	33÷45	65÷69	Satisfactory	C+	Satisfactory knowledge of subject, key aspects of area of study, satisfactory degree of skills and experience in practical activities. Learning outcomes are achieved however some of them are evaluated within the minimum mark range for the course.
			55÷64		C	
0% - 54%	22÷40	33÷60	55÷100	Passed	D	Learning outcomes are evaluated within the minimum mark range for the course.
	0÷21	0÷32	0÷54	Non satisfactory / Failed	F	Learning outcomes are evaluated below the minimum mark range for the course.
0% - 54%	22÷40	33÷60	55÷100	Passed	D	Learning outcomes are evaluated within the minimum mark range for the course.
	0÷21	0÷32	0÷54	Non satisfactory / Failed	F	Learning outcomes are evaluated below the minimum mark range for the course.

CONCLUSION

This work is devoted to the preparation of elements of educational and methodological support of the discipline "Professional training in English" for students of the Engineering School of natural resources Department of Geology, studying in the direction of 05.06.01 Earth Science profile "land Management, cadastre and land monitoring" (Level of higher education training of scientific and pedagogical personnel in graduate school

The program is designed for a three-year course for students of recruitment in 2020. The discipline is taught in the second year (3-4 semesters). The paper presents the working program of the discipline, examples of evaluation tools, types of training activities and selected forms of organization of the process.

The paper gives an example of an educational and methodical complex of a similar discipline used in our universities.

The developed elements of EMCD are planned for partial implementation in the educational process starting from the autumn semester of 2020-2021.

ACKNOWLEDGMENT

This research received no grant from any organization.

REFERENCES

Bespalenko, O. N. (2012). Assessment of restoration dynamics of burnt forest area vegetation using Landsat satellite data. In Modern problems of land management, cadastre and land monitoring: textbook. VGLTA.

CV and Cover Letter Advice. (2023). https://www.jobs.ac.uk/careers-advice/cv-templates

Federal State Educational Standard of Higher Education in the areas of training "41.03.05. International relations", "41.03.01. Foreign Regional Studies". (2023). https://eng.kpfu.ru/bachelor-specialist-programs/arts-and-humanities-bachelor-programs/

Google translate. (2023). https://translate.google.com/?hl=ru

How to write a CV. (2023). https://www.prospects.ac.uk/careers-advice/cvs-and-cover-letters/how-to-write-a-cv

Kapterev, A. A. (2012). *Presentation skill. How to create presentations that can change the world from English*. Mann, Ivanov and Ferber, Eksmo.

Khalilova, L.A. (2007). *Foreign language in a non-linguistic University: what, why and how to teach*. GEOTAR-Media.

Lebedeva, N. A. (2022). Teaching of Foreign Language in the Educational Process of Public Administration Personnel in Ukraine. In O. Kulaç, C. Babaoğlu, & E. Akman (Eds.), *Public Affairs Education and Training in the 21st Century* (pp. 294–310). IGI Global. doi:10.4018/978-1-7998-8243-5.ch019

Murzin, A. D. (2013). *Real estate: Economics, valuation and development*. Phoenix.

Pasko, O., Kovyazin, V., Zakharchenko, A., & Lebedeva, N. (2020). Spatial organization of forestfund. *IOP Conference Series. Earth and Environmental Science*, *574*(1), 012061. doi:10.1088/1755-1315/574/1/012061

Pasko, O., Staurskaya, N., Zakharchenko, A., Zharnikov, V., & Larionov, Y. (2022). *Bio-Farming as the Basis of Environmentally-Sustainable Arable Farming at the Time of Global Warming*. Research Anthology on Strategies for Achieving Agricultural Sustainability. doi:10.4018/978-1-6684-5352-0.ch031

Pasko, O., Zakharchenko, A., & Kovyazin, V. (2021). *Analysis of land management of the forest fund*. Bulletin of the Tomsk Polytechnic University. *Geo Assets Engineering.*, *332*(2), 127–138.

Pasko, O. A., & Kovyazin, V. F. (2017). *Research work of a master student: teaching aid*. TPU.

Polat, E. S., & Bukharkina, M. Yu. (2007). Modern pedagogical and information technologies in the education system. Ed. Center "Academy".

Senko, Yu. V., & Frolovskaya, M. N. (2007). *Pedagogy of understanding*. Drofa.

Shmatko, N., & Volkova, G. (2023). *Knowledge of foreign languages by Russian scientists*. https://issek.hse.ru/news/226574939.html

Sinyansky, I. A. (2014). *Typology of real estate: textbook for universities*. Academy.

Slezko, V. V. (2013). *Land Management and land use management*. Infra-M. doi:10.12737/1680

TemplatesR. (2023). https://uptowork.com/resume-templates#1

Tokareva, O. S., Alshaibi, A. D. A., & Pasko, O. A. (2021). *Assessment of restoration dynamics of burnt forest area vegetation using landsat satellite data*. Bulletin of the Tomsk Polytechnic University. *Geo Assets Engineering.*, *332*(7), 191–199.

Transcription, pronunciation and translation of English words online. (2023). https://myefe.ru/anglijskaya-transkriptsiya.html

Use Harvard Referencing Style. (2023). https://curve.coventry.ac.uk/open/file/bdfb947c-9d43-48d3-8ec8-f511682e1dd1/1/The%20CU%20Guide%20to%20Referencing%20in%20Harvard%20Style.pdf

Van Dijk, T. A. (1989). *Language. Cognition. Communication.* Progress.

Varlamov, A. A., & Galchenko, S. A. (2012). *State real estate cadastre.* Koloss.

Vitae, C. (CV) Samples, Templates and Writing Tips (2023). https://zety.com/blog/cv-format

Walger, J. W. (2015). Real estate as an object of capital investment. Problems of market economy management interregional collection of scientific works. Tomsk Polytechnic University (TPU).

APPENDIX

Example of Syllabus on a Similar Topic for Further Use

Land Use Planning and Control Syllabus

Law 6460, Section 0220, 3 credits
Meeting time: 8:30 AM to 9:20 AM, Fall 20212
Meeting days: Monday, Tuesday, and Wednesday
Meeting room: 355C
Instructor: Thomas Hawkins
Instructor contact: wthomas@ufl.edu, (352) 377-3141

Office Hours

As adjunct faculty, I do not maintain regular office hours at the law school. I am available to discuss the course material after each class period. I would also be happy to schedule a meeting at a time convenient to you. You may call or text me during daytime hours seven days a week at (352) 377-3141.

Required Text

The course supplement, titled Land Use Law with a Florida Focus LAW 6460 Course Supplement, Fall 2021, is available for purchase in the law school bookstore.

Additional materials will be posted on TWEN throughout the semester.

Course Description

A study of the legal aspects of the allocation and development of land resources; private controls through covenants and easements; public regulation and control through zoning and subdivision regulation; social, economic and political implications of land regulations; eminent domain; selected current problems such as growth management, historic preservation, environmental regulations, and

Purpose of Course

Land Use Planning and Control will give students a working knowledge of (1) the basis for state and local powers to regulate land use, (2) constitutional limits on state and local government powers including the restraints imposed by the Takings Clause, the Equal Protection Clause and Substantive Due Process, (3) the impact of Procedural Due Process on the standards of review for growth management decisions, (4) the regulatory framework for growth management established by Florida Statutes and (5) tools for funding the public and private infrastructure that development requires.

Course Goal

Students will apply land use law from Florida Statutes, Florida Courts and Federal Courts to hypothetical problems to describe the rights of property owners, to describe the rights of other parties affected by uses of land and to evaluate likely outcomes.

Course Objectives

Students will read cases and statutes to become familiar with judicial analyses of land use conflicts. Students will attend class and engage in class discussions to learn from the instructor and peer discussion of assigned cases and statutes. Students will attend quasi-judicial hearings and write a description of the hearing and a comparison of their understanding of due process protections in quasi-judicial hearings to actual practice.

Assessment

I will assign you a course grade based on the following:

1 Preparedness for, and participation in, class discussion (10% of final grade).
2 Attendance at local government quasi-judicial hearing (20% of final grade)

During the semester you will attend a local government quasi-judicial hearing and write a response paper documenting your attendance and impressions.

3 Final Exam (70% of final grade)

The take-home final exam will be administered on December 13, 2021

Course Schedule of Topics and Assignments

This course will address the following topics in the order presented. This order of materials corresponds to the order these topics are addressed in the course supplement. Assignment due dates will be announced during the semester. To prepare for each class session, read approximately 30 pages in the course supplement or assigned additional material beyond what has been previously covered in class discussion.

I. INTRODUCTION
II. NUISANCE
III. LOCAL GOVERNMENT LAND USE POWERS
IV. PROCEDURAL DUE PROCESS
   A. The Fairly Debatable Standard
   B. The Strict Scrutiny Standard
V. LAND USE DECISIONS IN FLORIDA
   A. Statutory mandate for planning
   B. Comprehensive planning decisions are legislative
   C. Process required in land use hearings
   D. Particular quasi-judicial decisions
   E. Challenging local government land use decisions
VI. THE TAKINGS CLAUSE
   A. Eminent domain
   B. Regulatory takings
   C. Exactions
VII. STATE PRIVATE PROPERTY RIGHTS PROTECTION
   A. Equitable estoppel
   B. Harris Act
VIII. PAYING FOR INFRASTRUCTURE
   A. Impact fees
   B. Tax increment financing
   C. Community Development Districts
   D. Concurrency
   E. Development Agreements
   F. Contract zoning
IX. EQUAL PROTECTION
   A. Class-based discrimination
   B. Classes of one
X. SUBSTANTIVE DUE PROCESS
   A. Free speech
   B. Free exercise and the RLUIPA
XI. OTHER U.S. CONSTITUTIONAL LIMITATIONS ON LAND USE REGULATION
   A. Establishment Clause
   B. Dormant Commerce Clause

Class Preparation and Workload

To prepare for each class session, spend approximately two hours studying assigned readings and preparing responses to hypothetical prompts. Policy related to make-up exams or other work.

This course follows the law school policy on delays in taking exams.

Statement Related to Accommodations for Students With Disabilities

Students requesting classroom accommodation must first register with the Office of Disability

Resources. The UF Office of Disability Resources will provide documentation to the student who must then provide this documentation to the Law School Office of Student Affairs when requesting accommodation.

University Policy on Academic Misconduct

UF students are bound by The Honor Pledge which states, “We, the members of the University of Florida community, pledge to hold ourselves and our peers to the highest standards of honour and integrity by abiding by the Honor Code. On all work submitted for credit by students at the University of Florida, the following pledge is either required or implied: “On my honour, I have neither given nor received unauthorized aid in doing this assignment.” The Honor Code (https://www.dso.ufl.edu/sccr/ process/student-conduct-honour- code/) specifies the number of behaviours that violate this code and the possible sanctions. Furthermore, you are obligated to report any condition that facilitates academic misconduct to appropriate personnel. If you have any questions or concerns, please consult with the instructor in this class.

Information on UF Grading Policies

Table 11. This course follows the law school grading policy, letter grades earn the following points

Letter Grade	Point Equivalent
A (Excellent)	4
A-	3.67
B+	3.33
B	3
B-	2.67
C+	2.33
C (Satisfactory)	2
C-	1.67
D+	1.33
D (Poor)	1
D-	0.67
E (Failure)	0

The Online Course Evaluation Process

Students are expected to provide feedback on the quality of instruction in this course by completing online evaluations. Evaluations are typically open during the last two or three weeks of the semester, but students will be given specific times when they are open. The summary results of these assessments will be available to students.

Attendance Policy

Requirements for class attendance and make-up exams, assignments, and other work in this course are consistent with university policies that can be found at: https://catalog.ufl.edu/ugrad/current/ regulations/info/attendance. Additionally, according to ABA guidance, absences from more than 15% of classes will render a student ineligible to receive credit for the course.

TASK 7. PREPARE DEMONSTRATION MATERIALS FOR CONDUCTING CLASSES IN THE DISCIPLINE.

Demonstration materials allow you to visualize information and facilitate its assimilation (Fig.2). Traditionally, classes are accompanied by presentations, and recommendations for creating which can be found in (Pasko, Kovyazin, 2017; Kapterev, 2012).

Figure 2. Examples of presentations on the topic "monitoring of agricultural land"

As an example, let to cite the author's unique experience gained during a visit to the Democratic People's Republic of Korea as part of a delegation at the invitation of the country's government (Fig. 3). The Pedagogical Institute showed the latest demonstration materials addressed to future teachers of the English language. According to the Korean concept of learning, sense organs should be used to the maximum for understanding and memorization.

Figure 3. The author during a visit to North Korea. A) media interviews; B) on the territory of the National Pedagogical University. Photo by the author of this chapter – Olga Pasko

A B

For example, guests were shown numerous albums with photographs of animal habitats and the animals themselves. In envelopes, we saw a colorful album about the bird, in which the envelopes contain grains that it feeds on, feathers from the nest, collected after molting. By pressing a special button, you can hear the voice of birds, and by placing a CD, you can see a mini-movie. Samples of nests, eggs, stuffed birds are stored in the office (Figure 4).

The use of the latest computer technologies in the English class was unusual. With their help, the teacher can choose his image (a famous person, a fairy-tale character, etc.) and the language of communication.

The teacher led the lesson (moved, spoke), and on the plasma screen the same words and movements were repeated by scientists, artists, fairy-tale characters in the appropriate voice. This form of conducting classes is very popular among students.

Korean inventors also showed a creative approach in making 3-D models of the mountain peaks of the planet.

Figure 4. All the senses should be involved in the learning process – Korean teachers believe. Photo by the author of this chapter – Olga Pasko.

With the help of sand and a projector with the image of a difference in heights, certain mountain ranges were to be formed on the surface and the accuracy of the task was to be checked according to their profile. This work is completed by its text description in the report (Figure 5).

Figure 5. Students and teachers form a 3-D relief of the mountain peaks of Korea from sand. The work is accompanied by an active discussion in a foreign language. Photo by the author of this chapter – Olga Pasko.

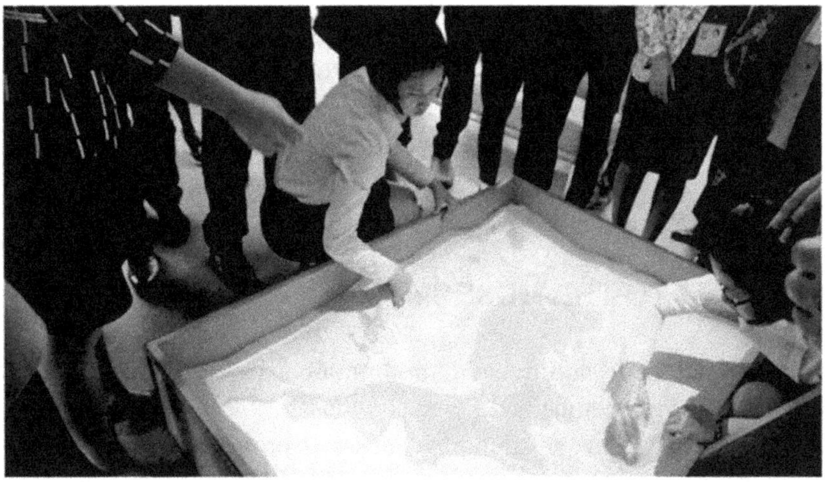

Since communication involves knowledge of not only a professional foreign language, the university has rooms with a family atmosphere (Figure 6).

Figure 6. To develop knowledge of a foreign language at the household level, the National Pedagogical University of North Korea has rooms resembling apartments. Photo by the author of this chapter – Olga Pasko.

Playfully, again with the use of the latest electronic devices, pedagogical skills are assessed. Various animated images (class, expedition, exam, etc.) appear before the student on a large, wall-to-wall plasma screen. Suddenly, the situation changes dramatically, for example, a bird flies through the window. The class is distracted, animated and out of control. What should the teacher do (his professionalism is assessed by senior colleagues)? In the case of the bird, he suggested that the students take their seats so that the bird would calm down and find the open window on its own. "Let's not disturb her. And by the way, what bird is depicted on the coat of arms of our country? Eagle – is the students' answer. It is right. Please, write what is remarkable about this bird.

TASK 8. VISIT TO THE MUSEUM

Conducting methods: audiovisual, communicative, reproductive-creative, suggestopedic and project method.

The task is to create an author's "museum project", which includes the implementation of creative and educational tasks (collection of information and illustrative materials, preparation of a report, presentation, photo essay, video, stand, preparation and conduct of role-playing games and quests, etc.).

In the thematic halls, both natural and virtual exhibits coexist, the authenticity of which strikes the imagination and arouses research interest.

Figure 7. Entrance to the Museum of Natural History in Pyongyang and expositions of the origin and development of life on Earth. Photo by the author of this chapter – Olga Pasko.

TASK 9. IMPLEMENTATION OF THE AUTHOR'S PROJECT

This task does not directly relate to the topic of advanced training for teachers of non-linguistic specialities, but it is extremely valuable for any education at its transcendental peak. This is an annual competition of architectural projects among students of universities in North Korea.

Its winners get a fantastic opportunity to implement the project with any, of the most daring ideas. This becomes the task of the authorities of Korea for the builders, in fact, the law.

Figure 8. Pyongyang's New Quarters Based on Designs Winning Architectural Competitions by Students of Korean Universities. Photo by the author of this chapter – Olga Pasko.

Thus, the process of teaching English to teachers of non-linguistic specialties is a complex, creative, dynamic process. It requires not only fluency in the language but also knowledge of the basics of the specialty and psychology.

Chapter 7
Improving the Lexical and Grammatical Skills of Students Specializing in Hydraulic Engineering and Water Resources by Means of Writing

Nadezhda Anatolievna Lebedeva
Kherson State Agrarian and Economic University, Ukraine

EXECUTIVE SUMMARY

Writing plays a very important role in the life of every technical sphere specialist. The purpose of this chapter is to describe organizational and methodical writing bases and improving the lexical and grammatical skills of students specializing in hydraulic engineering and water resources by their means. Achieving this goal involves the following task to solve: consider the main methods and techniques of teaching a foreign language to students of the specialty Hydraulic Engineering (Water Resources). The scientific novelty of the results obtained lies in the fact that, although the chapter is descriptive, theoretical in nature, it may be interesting to exchange experience with colleagues from other countries. The results of the study can be used to develop methodological manuals, innovative means of teaching foreign languages for professional communication, and advanced training for research and teaching staff.

DOI: 10.4018/978-1-6684-6222-5.ch007

INTRODUCTION

Writing plays a very important role in the life of every technical sphere specialist. Writing is recognized as the most complex type of speech activity, which involves all speech analyzers. The material base of writing consists of spelling and graphic skills. Writing is an expressive productive type of speech communication. The process of written speech, as we know, begins with internal speech. Such kind of it is that a program of utterance is prepared by the selection of lexical units, the grammatical design of sentences, and their logical combination in paragraphs. The whole process of writing consists of internal speaking and recording of the prepared material on paper, which requires the automated operation of sound-graphic associations. In terms of the number of operations, a written message is more complicated than an oral one. However, in oral spontaneous expression, the generation of speech and its sound design is synchronous and therefore requires full automation in the operation of lexical and grammatical material. The process of learning to write is facilitated by the fact that the writer does not feel the lack of time, and this allows him to consider the content and form of future statements more thoroughly, and more clearly carry out both preliminary synthesis and retrospective analysis of the writing. The ability to find in memory the necessary language tools for an accurate and clearer expression of thought, and the ability to use a dictionary and other reference books give the writer more confidence. The lexical and grammatical skills of students within the boundaries of a specialized subject will improve the culture of professional communication. They develop through practical writing exercises, in particular, translation from a foreign language into a native language and vice versa.

The purpose of this chapter is to describe organizational and methodical writing bases and improving the lexical and grammatical skills of students specializing in hydraulic engineering and water resources by their means.

Achieving this goal involves the following tasks to solve:

1. Consider the main methods and techniques of teaching a foreign language to students of the speciality “Hydraulic Engineering (Water Resources)”.
2. Outline the linguo didactic foundations of teaching English-language written communication in hydraulic engineering.
3. Determine the content of teaching English-language writing and determine the criteria for selecting educational material for the formation of competence in English-language written communication on Hydraulic Engineering (Water Resources).

The object of research is the process of teaching professionally oriented English-language written communication on hydraulic engineering and water resources in agricultural institutions of Higher Education in Ukraine.

The research subject is the methodology of teaching professionally oriented English – language written communication on hydraulic engineering and water resources.

The research hypothesis is that the effectiveness of teaching professionally oriented foreign language written communication with hydraulic engineering and water resources can be improved by increasing the lexical and grammatical skills of students specializing in hydraulic engineering and water resources by means of writing, which provide an expansion of the communicative and linguistic components of professional competence in foreign language written communication through the step-by-step use of the appropriate exercises and tasks system.

To solve the tasks set, the following methods were used: a theoretical and critical analysis of domestic and foreign scientific literature in the pedagogical field and methods of teaching foreign languages, which made it possible to study and generalize modern methods of teaching professional foreign language writing, systematize and clarify the theoretical foundations of solving the problem; analysis of textbooks and programs in a foreign language in a professional direction; forms and methods of teaching professionally oriented English-language written communication; empirical observations of the educational process, students ' activities, a means of checking the effectiveness of written tasks on hydraulic engineering and water resources. The scientific novelty of the results obtained lies in the fact that, although the chapter is descriptive, theoretical in nature. However, it may be interesting to exchange experience with colleagues from other countries. The results of the study can be used to develop methodological manuals, innovative means of teaching foreign languages for professional communication, and advanced training for research and teaching staff.

BACKGROUND

Modern society places high demands on the quality of training of specialists in the agricultural sector of the economy, in particular, specialists in hydraulic engineering and water resources, who, according to international standards, "must have the scientific foundations of their business, be guided by professional problems, concepts and terms, be fluent in a foreign language at a professional level, be creative in their activities, continuously replenish their own knowledge, as well as from foreign sources, be ready to study and work in any country in the world. The process of integrating Ukrainian education into the European educational space increases the importance of learning a foreign language not only as an academic discipline in higher education institutions but also as a means of mastering a specialization that is consistent with the requirements of the Bologna declaration and the lingua didactic policy defined by the recommendations of the Council of Europe Committee on Education" (Zhovnich, 2018, p. 1). Therefore, the training of future specialists in the field of hydraulic engineering and water resources who are fluent in a foreign language of professional activities is an important task of modern higher education. At the same time, the declaration of European principles for the formation of foreign language competence of future specialists who will work in a multilingual, United World requires improving the methods of teaching a foreign language. In the working program "Foreign language for professional purposes" for applications for higher education, studying in the educational program "Bachelor" in the speciality "6.060103. Hydraulic Engineering (Water Resources)" (Lebid, Lebedeva, 2018), there is a purpose of studying the discipline. It is to form the necessary communication ability in the areas of professional and situational communication in oral and written forms, skills of practical command of a foreign language in various types of speech activity in the volume of topics that are determined by professional needs; mastering the latest professional information through foreign sources.

The task of the discipline is to teach, develop and improve various types of speech activity, listening, speaking, dialogic speech, reading, writing and translation. Following Ukrainian educational standards and training in various types of speech, the activity includes the assimilation of:

- phonetic norms of a foreign language;
- 2000 lexical units, knowledge of which provides students with the opportunity to conduct a conversation on the speciality and get information from foreign written and oral sources;
- grammatical material of a foreign language;
- lexical minimum (categories of being, their properties and relations; geographical, demographic, economic and political data) of the specific country of the world whose language is being studied;
- abbreviations of foreign-language professional terms in a certain professionally-oriented industry;
- lexical minimum of business contacts, business meetings and so on;
- speech etiquette of communication, speech models of address, politeness, apologies and approval;
- lexical and grammatical minimum ensuring the language and communicative level of presentations;
- linguistic methods of analytical processing of foreign language sources;
- working with electronic foreign-language sources;
- lexical minimum of computer technologies (Lebid, Lebedeva, 2018).

Learning to write contributes to the formation of speaking and reading skills, so for this type of speech activity, it is determined that it is appropriate to teach students not only spelling, but also the ability to express their thoughts in writing, and at the same time involves mastering:

- foreign-language features of business correspondence; vocabulary, grammar, the syntax of business etiquette and cultural aspect of written communication;
- methods and linguistic features of annotating and referencing foreign language sources (Lebid and Lebedeva, 2018).

It is well known that writing is a powerful factor in the accelerated development of mankind. It plays an important role in the creation and preservation of spiritual, and through them, material values. The letter is international. It continues and develops everything that was created by other people. Writing is not a natural phenomenon, but a social phenomenon, an invention of Man, which was used and is used by everyone.

Psycholinguistic Aspect of Writing

Writing is recognized as the most complex type of speech activity, the material base of which is spelling and graphic skills. Like speaking, writing is an expressive type of speech communication (Artemov, 1969; Atkinson, 1973).

As a means of forming and formulating thoughts, writing is based on the use of speech signs fixed by the nervous connections of the cerebral cortex in the form of visual and hand-motor images interacting with auditory and speech-motor ones. People who know how to read and write have images of words in addition to sound ones. Graphic images have hand-motor parallels that are controlled by the motor mechanism of the writing hand. Spelling images of words are fixed precisely in the speech, and not in the motor mechanism of the hand (Artemov, 1969; Atkinson, 1973).

Graphically correctly represent the letters of the alphabet, transform sounds and sound combinations into letters, while associating the sound, graphic and kinetic images of a word with its meaning; spell correctly write words, and phrases, put punctuation marks; operate with speech structures; be able to formulate your thoughts in written form - all this is combined in the ability to write. Internal speech prevents the writing process. It is in it that the program of utterance is prepared. In an internal speech, lexical units are selected and sentences are designed grammatically. Logical combination in paragraphs also occurs in internal speech. According to Bekh, Birkun (1996), Vishnevsky (1979), and Davydova (1990), the whole process consists of internal pronunciation and fixing the prepared material on paper. This result requires the automated operation of sound-graphic associations. In terms of the number of actions, writing is more complex than oral activity. However, in oral spontaneous utterance, the generation of speech and its sound design is synchronous and therefore requires complete automatism in the operation of lexical and grammatical material.

The process of learning to write is facilitated by the fact that the student who writes does not feel a lack of time, and this allows him to think through the content and form of the future statement regarding hydraulic engineering and water resources thoroughly, and more clearly carry out both synthesis and analysis of what is written. The ability to find in memory the necessary language tools for accurate and clear expression of thoughts on the topic, and the ability to use a dictionary and other reference books of specialized vocabulary on hydraulic engineering give students more confidence.

As established by scientists Davydova (1990), Fernos (2019), Shcherbina (2019), as a result of regular training in written utterance, students gradually gain experience, the process of choosing words and structures passes faster, and writing skills become more stable. The ability to express one's own and other people's thoughts in writing is important not only in itself but also as a means of improving the level of oral speech proficiency in the speciality. It is known that an oral message becomes more thorough and logical when it is prepared in writing. Written speech becomes a prerequisite for the development of internal speech. Writing is a conscious act. Using a foreign language further increases the degree of awareness. According to Gurvich (1990), Ustimenko (2004, a, b), and Folomkina (1991), the combination of auditory-speech-motor images of internal speech and visual-hand-motor images in the writing process provides comprehensive assimilation of knowledge. Stable associations created in this way contribute to more effective perception, memorization and reproduction of material from hydraulic engineering and Water Resources. The strength of memorization of specialized vocabulary is ensured by the participation of all analyzers, the interaction of analysis and synthesis, awareness and the involuntary writing process. Graphic images of words and structural models of sentences learned through written exercises are more easily recognized by visual perception during reading and translation. In the written speech of the speciality, the speech nature of a coherent utterance concerning hydraulic engineering is more clearly revealed, which usually consists of several logically and structurally related sentences.

RESULTS AND DISCUSSION

Exercises for Improving the Lexical and Grammatical Skills of Students

Writing is one of the most effective means of control and self-control for students and contributes to the formation of skills. A large amount of writing in foreign language homework is not accidental, as students attach more importance to completing written tasks than oral ones.

In classroom work in a foreign language for special purposes, writing is the most economical, reliable, effective and mass-produced way to control knowledge and the level of proficiency in the specialized vocabulary.

According to the selected stages of formation of writing skills and abilities, the system of exercises for teaching writing by Folomkin (1991) and the Council of Europe Education Committee (1998) includes the following groups of exercises:

- exercises for developing writing skills;
- exercises for developing speech writing skills;
- exercises for developing writing skills.

The purpose of teaching technical writing of hydraulic engineering and water resources is to form actions that ensure the presence of language material in writing. Typical types of exercises at this stage can be seen in the following figure:

Figure 1. Types of exercises in teaching technical writing and written speech (summarized by the author of the chapter)

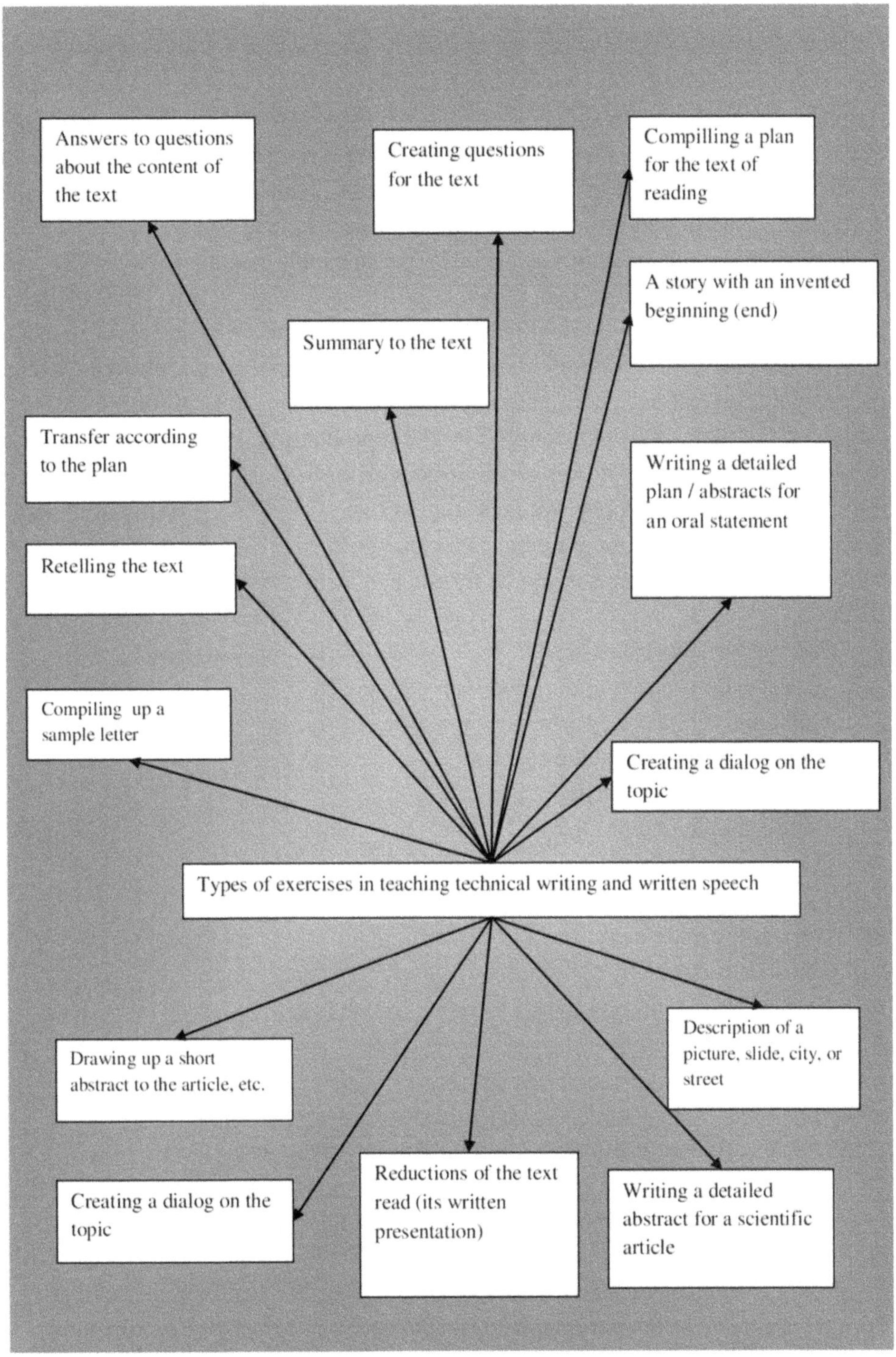

As it is noted already, two types of exercises are recommended for teaching written speech, that is, the ability to express oneself using graphic signs: conditional-communicative and communicative. Conditional communication exercises are used for mastering speech material in graphic design, as well as for automating operations related to the written presentation of thoughts. This type includes exercises with models (transforming models, expanding them, etc.). Exercises to expand the utterance are of particular importance. The value of such exercises is that they form a mechanism of advance or a mechanism of probable prediction, which is of great importance for creating a written utterance about hydraulic engineering and water resources.

The second type of training exercise provides the ability to logically and consistently convey this or that information in one's own words. This includes paraphrasing, shortening the text, and selecting keywords from the text. To teach the logical presentation of thoughts, it is proposed to draw up various plans: a plan in the form of questions or nominative sentences, and so on. Among communication exercises, the main place is occupied by the reproduction of groups of sentences related to the content. The exercise is performed in the form of self-dictation. To improve the ability to express yourself in your own words, use retelling, narration, description, essay, or writing a letter to a colleague.

Tasks for written exercises can be formulated as follows:

1) Answer the questions.
2) Ask questions about underlined words / phrases.
3) Combine several (2-3) sentences into one.
4) Rewrite and complete the sentences.
5) Choose the correct answer.
6) Report on the beginning / end of the story model.
7) Write the opposite.
8) Fill out the questionnaire by answering "yes" or "no".
9) Correct the statement.
10) Create a dialogue by combining disparate replicas.
11) Express your opinion about... by using a sentence...
12) Make a report on...
13) Write a resume.

To work with text, it can be offered the following tasks:

1) Write out sentences from the text that reveal their meaning.
2) Ask questions about the content of the text.
3) Make a summary of the text you read.
4) Give a name to the paragraphs of the text.

5) Write out new words and expressions.

It is also important to provide students with translation information. As Fernos (2021) notes, "despite the fact that in recent years, due to the reform of higher education in learning foreign languages, practical knowledge of a foreign language as a means of communication has come to the fore, the role of scientific and technical translation in the process of training certified specialists in the agricultural sector continues to remain even now. This in no way contradicts the requirement of teaching students oral communication, which is now very relevant in light of the requirements of the Bologna Process. When preparing a student to read literature in a narrow speciality, the teacher should clearly outline the lexical minimum required for the student to read this type of text, it is advisable to develop and publish dictionaries-minimums in specialities" (Fernos, 2021, P. 207).

Here is an example from the special edition "Methodological developments for the improvement of professionally-oriented communicative competence of students (based on English-language texts of the professional direction for the speciality "6.060103. Hydraulic Engineering (Water Resources)" by Sereda & Bokshan (2008), which was developed at the Department of foreign languages of the Kherson State Agrarian University, which students study nowadays:

Dams and Their Types

A dam is a barrier that divides waters. Dams generally serve the primary purpose of retaining water, while other structures such as floodgates, levees, and dikes are used to prevent water flow into specific land regions.

Dams can be formed by human agency, natural causes, or even by the intervention of wildlife such as beavers. Man-made dams are typically classified according to their size (height), intended purpose or structure.

By size

International standards define large dams as higher than 15 meters and major dams as over 150 meters in height.

By purpose

Intended purposes include providing water for irrigation or town or city water supply, improving navigation, creating a reservoir of water to supply industrial uses, generating power, creating recreation areas or habitat for fish and wildlife, flood control and containing effluent from industrial sites such as mines or factories.

A saddle dam is an auxiliary dam constructed to confine the reservoir created by a primary dam either to permit a higher water elevation and storage or to limit the extent of a reservoir for increased efficiency. An auxiliary dam is constructed in a low spot or saddle through which the reservoir would otherwise escape. On occasion, a reservoir is contained by a similar structure called a dike to prevent inundation of nearby land. Dikes are commonly used for reclamation of arable land from a shallow lake. This is similar to a levee, which is a wall or embankment built along a river or stream to protect adjacent land from flooding.

An overflow dam is designed to be over topped. A weir is a type of small overflow dam that can be used for flow measurement.

A check dam is a small dam designed to reduce flow velocity and control soil erosion. Conversely, a wing dam is a structure that only partly restricts a waterway, creating a faster channel that resists the accumulation of sediment.

A diversionary dam is a structure designed to divert all or a portion of the flow of a river from its natural course.

By structure

Based on structure and material used, dams are classified as timber dams, arch-gravity dams, embankment dams or masonry dams, with several subtypes. (Sereda & Bokshan, 2008, c. 3)

Vocabulary

floodgate - водоскид
large dam - високонапірна гребля; велика гребля (вище 15 м)
major dam - велика гребля
saddle - підкладка; подушка на вершині пілона висячого моста
auxiliary dam - допоміжна гребля
inundation - повінь, затоплення
reclamation - меліорація
embankment - насип, дамба, вал, гребля
embankment dam - земляна гребля

overflow dam - водоскидна гребля
weir - гребля, водоскид
check dam - низьконапірна гребля, захисна дамба
wing dam - сполучна дамба
diversionary dam - водозабірна (водопідйомна) гребля
timber dam - дерев'яна дамба
masonry dam - гребля з кам'яної кладки (Sereda, Bokshan, 2008, c. 4)

EXERCISES

1. Find the English equivalents of the words and word-combinations:

Дамба (гребля); людський фактор; високонапірна гребля; запобігти затопленю; зрошування орних земель; швидкість потоку; запобігати накопиченню осаду; дерев'яна дамба.

2. Odd one out:
   1. levee, floodgate, dam, dike;
   2. natural causes, flooding, human agency, intervention of wildlife;
   3. river, effluent, stream, arable land, lake;
   4. large dam, embankment dam (or masonry dam), timber dam, arch-gravity *dam;*

   5.check dam, major dam, diversionary dam, saddle dam, overfloe dam.
3. Match the parts of the sentences:
   1. Dams generally serve the 1. … in a low spot or saddle through primary purpose … which the reservoir would escape.
   2. Intended purposes 2. … timber dams, arch-gravity dams, include … embankment dams or masonry dams with several subtypes.
   3. An auxiliary dam is 3. … reclamation of arable land from constructed … a shallow lake.
   4. Dikes are commonly used 4. … of retaining water. for …
   5. Based on structure and material 5. … providing water for irrigation or used, dams are classified as … town or city water supply, improving navigation, creating a reservoir of water to supply industrial uses, generating power,

flood control etc. (Sereda, Bokshan, 2008, c. 4)

4. Complete the sentences with suitable words and word-combinations:
   1. A dam is a barrier that … … .
   2. Intended purposes include providing water for … or town or city water supply, improving …, creating … … to supply industrial uses, generating …, … recreation areas, … control and containing effluent from industrial sites.
   3. An auxiliary dam is constructed in … … or … through which the reservoir would otherwise escape.
   4. A weir is a type of small overflow dam that can be used for … … .
      5.A wing dam is a structure that only partly restricts …, creating a faster channel that resists … … .
   5. Answer the following questions:
      1. What is the primary purpose of dams?
      2. How can dams be formed?
      3. How can they be classified?
      4. What functions do intended purposes include?
      5. What is a saddle dam?
      6. A small overflow dam is constructed for flow measurement, isn't it?
      7. What is the difference between a check dam and a wing dam?
      8. Does a diversionary dam divert the flow of the river or reduce flow velocity?
      9. *What is the classification of dams according to the structure and material used?* (Sereda, Bokshan, 2008, c. 5)

The effectiveness of this selection of texts and exercises is also confirmed by Fernos (2021). The researcher writes that when selecting texts for reading, it is necessary to take into account professional knowledge since there is a certain discrepancy between the level of information contained in a foreign text and the level of competence of the student in the corresponding field of knowledge in their native language. This can interfere with understanding a foreign-language text and reduce interest in reading literature in the speciality. As a result, there will be a negative attitude to learning a foreign language in general.

Therefore, first of all, together with teachers of special departments, the most relevant and interesting problems in the specialty for a particular stage of training are determined, taking into account which thematic cycles of systemically related educational materials in a foreign language are compiled. You cannot start learning to read with texts that differ, for example, in the complexity of the content and have a large amount of unfamiliar vocabulary. As a rule, information in popular scientific professionally-oriented texts, in contrast to articles from abstract journals, monographs and patents, is quite simple, in vocabulary and syntax, unified terminologically. (Fernos, 2021, p. 209)

Written exercises after the text are aimed at consolidating terminology. As it can be seen, these are exercises for reverse translation, exercises for questions and answers. This method of working with professional vocabulary when performing written exercises is quite effective in the field of hydraulic engineering and water resources. It allows students to learn professional information, comprehend it and promote monologue speech in a professional direction of hydraulic engineering & water resources in a foreign language.

By Marchenko (2021) approval, students often feel overwhelmed when starting an academic writing assignment such as a research paper, report, thesis, etc.

Academic writing is a new subject taught in Ukraine. It should help the postgraduate students in writing their scientific works correctly, following all rules and regulations of English. Some of the students have many problems with their native Ukrainian language while writing academically, thus, the task of the lecture is to observe and solve two-way problem. (p. 234)

When teaching writing an abstract or article, a university teacher may suggest that students first carefully read the entire article and determine its main idea. Then set the purpose / purpose of the article and write out reference sentences from each paragraph. The next step is to write out the main provisions from the article, and summarize the opinion of the author of the article. The final stage will be to reread the text, find out what thoughts of the author require additional argumentation, and add the necessary sentences. Learning how to write an abstract is based on a pre-compiled plan of the source text, as well as on a series of consecutive tasks Figure 2:

- Specify the authors ‘ last name and initials;
- Write the title of the article;
- Specify the place of publication, publisher, Year of publication, issue of the magazine, the main topic;
- Find out the main idea of each of the paragraphs;
- Group paragraphs according to the main idea of the text;
- Formulate the main idea / idea of the text.

Despite the fact that in recent years, due to the reform of higher education in learning foreign languages, practical knowledge of a foreign language as a means of communication has come to the fore, the role of scientific and technical translation in the process of training certified specialists in the agricultural sector continues to remain even now. This in no way contradicts the requirement of teaching students oral communication, which is now very relevant in the light of the requirements of the Bologna Process. When preparing a student to read literature in a narrow specialty, the teacher should clearly outline the lexical minimum required for the student to read this type of text, it is advisable to develop and publish dictionaries-minimums in specialties. (Fernos, 2021, p. 207)

Relationship of Writing With Other Types of Speech Activity

Researchers Ganina & Chumakova (2015) and Grishkova (2015) point out that writing Thoughts has significant learning benefits. This process is slow compared to oral speech, which makes it possible to focus more attention on individual stages of its implementation, and this, in turn, creates conditions for the use of such speech forms that are not yet sufficiently automated and therefore cannot yet be used in oral communication. The use of these forms in written speech guarantees a freer operation of them in oral speech.

"In the context of broad international relations with other countries, it is important to communicate with foreign specialists, develop professional, business and personal contacts with foreign partners and colleagues, and read various publications in a foreign language. Demand will continue to grow in the coming years. This is due to the fact that the number of students who learn a foreign language for general communication is decreasing. Educational institutions are coming to an increasing conclusion about the expediency of studying a foreign language in their specialty. The purpose of teaching a foreign language in higher educational institutions is both to master a foreign language as a means of communication, and to acquire professionally directed foreign language competence for the successful implementation of further professional activities. The socio-economic situation that has developed in the country today opens up new opportunities for self-realization for future specialists in the agricultural sector. Today, the situation on the labor market is developing in such a way that more and more large agricultural and processing enterprises in Ukraine, created jointly with foreign partners, need highly qualified specialists who are able to work using the latest technologies. Thus, students of agricultural institutions have a unique opportunity not only to gain theoretical knowledge in their chosen specialty, but also to consolidate it in practice in leading agricultural enterprises of both our country and a number of foreign countries, with which universities cooperate fruitfully" (Shcherbina, 2019, p. 252).

As researchers Gurvich (1980), Marchenko (2019), Fernos (2019) point out, in teaching vocabulary, grammar, written fixation of material allows you to create additional associations that help to memorize lexical units, sentences, structural patterns; organize written exercises (classroom and home) that contribute to the development of skills in using lexical and grammatical phenomena; repeat the material that is recorded in students ' notebooks; organize oral exercises, relying on structural samples fixed in notebooks or on the blackboard, lexical elements.

From a study by Ganina and Chumakova (2015), it is clear that writing in general has a positive impact on the learning process. Students write down a significant part of the educational material and use these notes when memorizing, repeating, fixing; they get used to various types of written exercises, which is a necessary element of developing skills and abilities. In the process of writing, all types of sensations are supported, and all analyzers are activated: acoustic, speech-motor, visual, and hand-motor. Consider, for example, working on the text "Mason Dams" (Wednesday, Bokshan, 2008, pp. 6-8). Reading and translating is the first stage of professional vocabulary development in hydraulic engineering and water resources. Next, we provide words that the teacher will definitely take out for dictation are provided. All exercises are performed in writing, and then fixed in oral speech. So, the whole process of teaching students in the specialty of hydraulic engineering and water resources is based on writing. Underestimation of the role of writing in teaching foreign languages in a professional direction negatively affects the entire educational process. It is known that in 90% of people, the visual channel of information perception is several times stronger than the auditory one. This suggests that the volume and strength of the learned material perceived through the visual communication channel will be significantly higher. Insufficient attention to writing, written exercises in teaching foreign languages, the performance of which is associated with the functioning of a larger number of analyzers compared to oral speech, leads to impoverishment of the educational process, to faster forgetting, to the loss of skills and abilities acquired by significant efforts. So, here is the text in a professional direction:

Masonry Dams

In the arch dam, stability is obtained by a combination of arch and gravity action. If the upstream face is vertical the entire weight of the dam must be carried to the foundation by gravity, while the distribution of the normal hydrostatic pressure between vertical cantilever and arch action will depend upon the stiffness of the dam in a vertical and horizontal direction. When the upstream face is sloped the distribution is more complicated. The normal component of the weight of the arch ring may be taken by the arch action, while the normal hydrostatic pressure will be distributed as described above. For this type of dam, firm reliable supports at the abutments (either buttress or canyon side wall) are more important. The most desirable place for an arch dam is a narrow canyon with steep side walls composed of sound rock. The safety of an arch dam is dependent on the strength of the side wall abutments, hence not only should the arch be well seated on the side walls but also the character of the rock should be carefully inspected.

Two types of single-arch dams are in use, namely the constant-angle and the constant-radius dam. The constant-radius type employs the same face radius at all elevations of the dam, which means that as the channel grows narrower towards the bottom of the dam the central angle subtended by the face of the dam becomes smaller. Jones Falls Dam, in Canada, is a constant radius dam. In a constant-angle dam, also known as a variable radius dam, this subtended angle is kept a constant and the variation in distance between the abutments at various levels are taken care of by varying the radii. Constant-radius dams are much less common than constant-angle dams. Parker Dam is a constant-angle arch dam.

A similar type is the double-curvature or thin-shell dam. Wildhorse Dam near Mountain City, Nevada in the United States is an example of the type. This method of construction minimizes the amount of concrete necessary for construction but transmits large loads to the foundation and abutments. The appearance is similar to a single-arch dam but with a distinct vertical curvature to it as well lending it the vague appearance of a concave lens as viewed from downstream.

The multiple-arch dam consists of a number of single-arch dams with concrete buttresses as the supporting abutments. The multiple-arch dam does not require as many buttresses as the hollow gravity type, but requires good rock foundation because the buttress loads are heavy. (Sereda & Bokshan, 2008, c. 6)

Vocabulary

upstream face - напірна грань греблі
cantilever - консоль, кронштейн; укіс
arch action - розпір
arch ring - архивольт арки
abutment - контрфорс, опора
buttress wall - стіна з контрфорсами
constant-angle (arch) dam - арочна гребля змінного радіусу
curvature - вигин, скривлення

concave lens - увігнута лінза
downstream - низовий бік греблі
multiple-arch dam - багатоарочна гребля
hollow gravity dam - масивно-контрфорсна гребля (Sereda, Bokshan, 2008, c. 7).

EXERCISES

1. Find the English equivalents of the words and word-combinations:

Гідростатичний напір; сила архивольта арки; крута бокова стінка; протилежний кут; кількість бетону; перекласти вантаж на основу та контрфорси.

2. True or false?
   1. The distribution of the normal hydrostatic pressure between vertical cantilever and arch action will depend upon the stiffness of the dam in a vertical and horizontal direction.
   2. The most desirable place for an arch dam is a wide canyon with steep side walls composed of concrete.
   3. The character of the rock should not be carefully inspected for an arch dam construction.
   4. Constant-radius dams are much less common than constant-angle dams. (Середа, Бокшань, 2008, с. 7)
3. Match the parts of the sentences:
   1. If the upstream face is 1. … that as the abutments channel
      vertical the entire weight grows narrower toward the bottom
      of the dam must be carried of the dam the central angle
      to … subtended by the face of the dam
      becomes smaller.
   2. The safety of an arch dam 2. … transmits large loads to the
      is dependent on … foundation and abutments.
   3. The constant-radius type 3. … the strength of the side wall
      employs the same face abutments.
      radius at all elevations of
      the dam, which means …
   4. This method of double- 4. … a number of single-arch dams
      curvature dam construction with concrete buttresses as the
      minimizes the amount of supporting abutments.
      concrete necessary for

construction but …

5. The multiple-arch dam 5. … the foundation by gravity.
   consists of …
   (Sereda, Bokshan, 2008, c. 8).

4. Complete the sentences with suitable words and word-combinations:
   1. The normal component of the weight of the arch ring may be taken by … … .
   2. The most desirable place for an arch dam is … … with … side walls composed of … rock.
   3. The constant-radius type employs the same face radius at all … of the dam.
   4. … … … are much less common than constant-angle dams.
   5. The multiple-arch dam requires … … … because the buttress loads are heavy.
5. Answer the following questions:
   1. How is stability obtained in the arch dam?
   2. What must the entire weight of the dam be carried to if the upstream face is vertical?
   3. What is important for a masonry dam?
   4. What is the most desirable place for an arch dam?
   5. What does the safety of an arch dam depend on?
   6. What are the two types of single-arch dams? Describe them.
   7. What are the peculiarities of a double-curvature dam construction?
   8. How is a multiple-arch dam constructed? (Sereda, Bokshan, 2008, c. 8).

So, the use of writing in foreign language classes on narrowly specialized topics has many qualities necessary for oral speech. It needs more thorough thinking to ensure the requirements of correctness, clarity, clarity, presentation of thoughts on hydraulic engineering and water resources, because it is isolated from the interlocutor, from the situation of communication. The more clearly there is a requirement to be more reasonable, logical, and evidence-based. Gradually mastering the features of writing on hydraulic engineering and Water Resources, the student transfers them to a certain extent into oral speech, which makes the latter more perfect, and the student feels more confident.

Writing as a Means of Control in Teaching a Foreign Language for Hydraulic Engineering and Water Resources

Writing can play a significant role in teaching oral speech, reading, vocabulary, and grammar. In addition, an important function of writing in teaching foreign languages is that at all stages it can serve as an effective means of control. Written test papers, both current and final, are used to test knowledge of language material, practical listening skills, oral utterance, and reading. Current control works are carried out after mastering a certain portion of language material on hydraulic engineering and water resources – at least once within two weeks. They include one or two tasks and last from 5 to 15 minutes. Final written test papers are conducted once a semester in accordance with the general schedule. Such works involve checking the proficiency of various types of speech activity and can last 90 minutes. Written works are always carefully prepared. Unprepared control work on hydraulic engineering and water resources undermines the student's self-confidence, causes feelings of fear and uncertainty, which negatively affects the academic performance of the student and the class as a whole, and causes a large number of mistakes.

There are the following forms of written control works:

Figure 2. Control works for testing knowledge of lexical and grammatical language material (summarized by the author of the chapter)

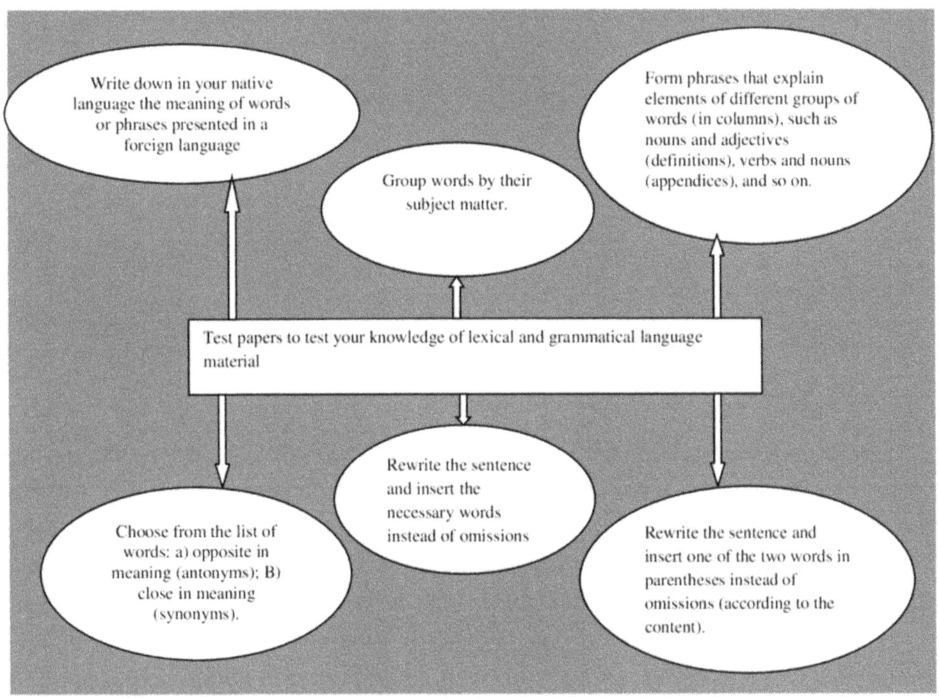

Knowledge of grammatical material can be tested by such tasks as writing these plural nouns, forming past tense forms from the submitted verbs, writing these verbs in an unassigned form, writing these adjectives to the highest degree, rewriting sentences, selecting and inserting the correct verb form (several forms are given in parentheses), etc. (Murphy, 1998; Bachov, 2006, Neil O'sullivan, James D.Libbin, 2011). The above tasks and exercises are purely conditional in nature. They can be supplemented with exercises for substitution, transformation, expansion, and answering questions. It is quite significant to control the ability to express yourself in writing, where the following options for tasks are possible:

Figure 3. Control of the ability to express yourself in writing (summarized by the author of the chapter)

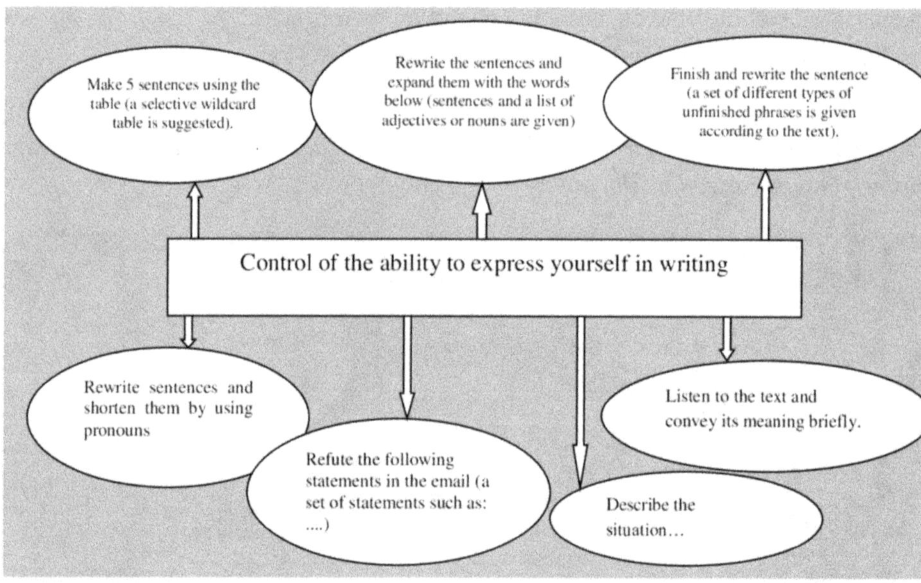

The form of presentation of the test paper may vary. Tasks can be dictated, written on the blackboard, projected onto the screen, or given on separate cards. It is useful to involve students themselves from time to time in checking current test papers by inviting them to check their own work or organize mutual verification of test papers using keys, a dictionary, a textbook, the correct version on the blackboard, and so on.

Summing up the above, we can say that writing is the purpose of teaching a foreign language in a professional direction, in particular in hydraulic engineering and water resources, and an effective means of teaching and controlling foreign language knowledge, skills and abilities. The psychophysiological basis of writing on hydraulic engineering and water resources is the interaction of auditory, hand-motor, speech-motor and visual analyzers. There is a distinction between graphic and spelling skills and written speech skills of hydraulic engineering and water resources. Such skills and abilities are formed through appropriate exercises. Among the main exercises that contribute to the development and improvement of creative writing skills of hydraulic engineering and Water Resources – essay, abstract, abstract, writing a thematic essay, etc.

Written translation from a foreign language into the native language and vice versa texts on hydraulic engineering and water resources contribute to the formation of students' skills of self-control. A large amount of writing in homework in a foreign language for professional purposes is not accidental. This fact is explained by the psychology of students: they attach more importance to written tasks than oral ones. In classroom work, writing is the most economical, reliable, effective and widespread way to control the knowledge and level of foreign language proficiency in the professional field.

CONCLUSION

As a result of regular training in writing, students who develop hydraulic engineering and water resources step by step gains knowledge. The process of acquiring special words and structures passes more quickly, and the skills of writing become stronger. The memory of the written expression of one's thoughts is important not only in itself but also as a tribute to the advancement of the soul's thoughts. The use of auditory-movement images of internal movement and visual-hand-motor images in the process of writing ensures the complex acquisition of knowledge of hydraulic engineering and water resources.

Students are invited to conduct a comparative evaluation analysis (independently or under the guidance of a teacher) of native and foreign texts. The texts are accompanied by a lexical-stylistic and translation-methodological commentary, which gives the rationale for the proposed translation, gives options that are acceptable in this text and contains appropriate references to the theoretical part (where necessary). Written translation of a technical text on irrigation and drainage is a conscious act from the very beginning.

All elements, forms, and tools used by students studying hydraulic engineering and water resources, through written translation, convey the desired content, which is the subject of his awareness. The use of lexical units of specialized topics further increases the level of awareness. The combination of auditory-motor images of internal speech and visual-motor images in the writing process provides a comprehensive acquisition of knowledge. The racks of the association created in this way promote more effective perception, memorization and reproduction of specialized technical material. The strength of memorization is ensured by the participation of all analysts, the interaction of analysis and synthesis, and the awareness and involuntariness of the translation process. Graphic images of words and structural models of sentences learned through written exercises are easier to recognize in visual perception. In a written speech, the speech nature of coherent utterance is more clearly revealed, which usually consists of several logically and structurally interconnected sentences.

So, in the process of teaching foreign language communication to students on hydraulic engineering and water resources, an important function of writing is that it can serve as an effective means of control. Written works, both current and final, are used to test knowledge of language material, and practical skills of listening, speaking and reading. Current tests with the use of technical translation are carried out after mastering a certain portion of language material or oral topic - at least once every two weeks. They include one or two tasks and last up to 15 minutes. Final written tests are conducted three times a semester following the curriculum. Such work involves testing the mastery of various types of speech activities of students studying hydraulic engineering and water resources and may last 90 minutes. So, writing is an important means of improving the lexical and grammatical skills of students specializing in hydraulic engineering and water resources. The main aim of this chapter is gained and all the tasks are solved by the author. Special texts and exercises have been used in the teaching process and approved by the students as useful and interesting.

ACKNOWLEDGMENT

This research received no grant from any organization.

REFERENCES

Artemov, V. A. (1969). *Psychology of teaching foreign languages*. Enlightenment.

Atkinson, R. (1979). *Human memory and the learning process*. Enlightenment.

Bakhov, I. S. (2006). *English Grammar Practice workshop on English grammar (2nd ed.)*. IAPM Publ.

Bekh, P. A., Birkun L. V. (1996). *Kontseptsiya ucheniya foreign languages V Ukrainy* [The concept of teaching foreign languages in Ukraine]. Academic Press.

Davydova, M. A. (1990). Activity-based methods of teaching foreign languages. Higher School.

Fernos, J. (2019). Some aspects of teaching terminological vocabulary during the course "foreign language (in a professional direction)" in a higher education institution. In *The III international scientific and practical Internet conference "Foreign language in professional training of specialists: issues and strategies"*. Kropyvnytskyi: EPC of Volodymyr Vynnychenko Central Ukrainian State Pedagogical University.

Folomkina, S. K. (1991). *Standard foreign language program for non-linguistic specialties. Foreign Languages in High School, 24.*

Ganina, E. V., & Chumakova, A. N. (2015). *Language problems in the global world.* Publishing House "Prospect".

Grabovska, N. A. (2021). Formation of digital reading skills in English in students of non-language specialists. Modern aspects of science: V-th volume of the collective monograph, 163-173.

Grishkova, R. O. (2015). *Methods of teaching English in the professional direction of students of non-philological specialties*. Mykolaiv.

Gritsik, N.V. (2015). Technology podcasting in teaching a foreign language in a professional direction. *Bulletin of Chernihiv National Pedagogical University. Series: Pedagogical sciences*, *124*, 24-26.

Gurvich, P. B. (1980). *Theory and practice of experiment in the methodology of teaching foreign languages*. Vladimir: Vladimirsk. state Pedagogical Institute.

Lebid, O. M., & Lebedeva, N. A. (2018). *Working program "Foreign language for professional purposes" for applications for higher education, studying in the educational program "Bachelor" in the specialty "6.060103. Hydraulic Engineering (Water Resources)"*. KSAU, Ukraine.

Marchenko, T. A. (2019). Academic writing for Ukrainian Ph.D. Students. In *The III International scientific and practical Internet conference "Foreing language in professional training of specialists: issues and strategies"*. Kropyvnytskyi: EPC of Volodymyr Vynnychenko Central Ukrainian State Pedagogical University.

Modern Languages: Learning, Teaching, Assessment. A Common European Framework of Reference (1998). Council of Europe. Education Committee.

Murphy, R. (1998). *English Grammar in Use*. Cambridge University Press.

Nikolaeva, S. Y., & Lenvi, K. (2005). English for specific purposes (ESP). *National Curriculum for Universities, 119.*

Nikolaeva, S. Yu., & Solovey, M. I. (2001). English Language Program for Universities / Institutes (five-year course of study). Project (Curriculum for English language Development in Universities and Institutes Moscow: British Council Publ.

O'Sullivan, N., & Libbin, J. D. (2011). *Agriculture.* Career Paths by Express Publishing.

Shcherbina, Y. (2019). Development of foreign language competence of agricultural students as a guarantee of their successful implementation in their professional activities. In *The III international scientific and practical Internet conference "Foreing language in professional training of specialists: issues and strategies"*. Kropyvnytskyi: EPC of Volodymyr Vynnychenko Central Ukrainian State Pedagogical University.

Ustimenko, O. M. (2004a). *Business English: curriculum.* Moscow: Publishing Center of the National Academy of Management.

Ustimenko, O. M. (2004b). *Independent work with English-language scientific and professional literature: methodological recommendations.* Moscow: Publishing Center of the National Academy of Management.

Verba, L. G., & Verba, G. V. (1997). *Grammar of modern English.* Logos.

Vishnevsky, O. I. (1982). Handbook of Foreign Language teachers. Moscow: Sovetskaya Shkola Publ.

Vishnevsky, O. S. (1979). *Substitution exercises in English.* Moscow: Sovetskaya Shkola Publ.

Zhovnych, O. V. (2018). *Methodology of Teaching Professionally Oriented English Writing to Future Journalists by Means of Blog Technologies* [Thesis]. Taras Shevchenko National University of Kyiv, Ministry of Education and Science of Ukraine.

Chapter 8
ESP Writing as a Practical Mean of Cultural Communication Between Ukrainian Music Teachers and Chinese Students

Zinaida Lebiedieva
Ukrainian National Tchaikovsky Academy of Music, Kyiv, Ukraine

EXECUTIVE SUMMARY

The purpose of this chapter is to present a practical experience of intercultural communication between Ukrainian music teachers and students from China using ESP through writing. During the study, the following tasks were set: 1) to study the literature on ESP for musicians, 2) conduct a survey of students from China using a questionnaire to determine the most convenient and intelligible form of communication, 3) analyze the received data and recommend the use of the ESP letter for Ukrainian and Chinese musicians. When researching, it is important to take into account the socio-cultural characteristics of Chinese students that affect their success in studying at higher musical institutions in Ukraine, revealing the peculiarities of the use of writing in the process of communication with teachers.

DOI: 10.4018/978-1-6684-6222-5.ch008

INTRODUCTION

Highly specialized English is of particular importance in terms of intercultural communication: it can act as a means of communication between Ukrainian teachers of the Academy of Music and Chinese students studying musical art since it is a foreign language studied both in Ukraine and in other countries. The topic of communication of the problem posed is especially relevant in the context of post-COVID reality, when many Chinese students continue their studies at Ukrainian musical institutions of higher education online, without being in the language environment of Ukraine, which entails certain difficulties in learning. To solve this problem, writing becomes an important tool in cultural and pedagogical communication. With the help of a specialized letter in English, a Ukrainian teacher and a Chinese student communicate. In a scientific paper Hu, Zh. (2013). He writes that in modern Chinese society, there is a change and development of intercultural, social, economic and political contacts between countries, and educational boundaries are expanding. "Over the past decades, cooperation between Ukraine and China in the educational field has achieved significant success, which has led to an increase in the interest of Chinese youth in higher education. In China, the prevailing opinion is that only education can provide a person with a worthy place in society, and thus a diploma of higher education allows you to make a career in the good sense of the word, becomes a pass to a better life" (Hu, 2013, p. 36). In recent years, despite the problems caused by the pandemic and the political situation in Ukraine, the number of Chinese students wishing to study in this country has increased. As you know, an important factor in the choice of Chinese students to study at Ukrainian universities is tuition fees. Compared to the UK or the US, studying in Ukraine is much cheaper. In addition, when entering universities in English-speaking countries, Chinese students must pass an English exam at a sufficiently high level. In Ukraine, the procedure for admission to Ukrainian universities for foreign students is more focused on subjects related to the speciality, that is, more simplified in terms of English proficiency. Despite the interest in receiving higher musical education in Ukraine, students from China face many difficulties caused by a foreign language and socio-cultural environment. In this case, English manifests itself as a means of intercultural communication. Moreover, given the various linguistic approaches to the typology of languages, the relevance of written communication is the most relevant.

In China, the state policy in the field of education, as noted by Yamin (2017), outlined several goals in the context of which a change in priorities in the system of pedagogical values leads to changes in the value-target guidelines and content of education. Because a large number of students from China study in Ukrainian universities, special attention is paid to the specifics of working with foreign students, and the features of musical and pedagogical education are studied. From the scientific work of Yamin (2017), it becomes clear that music education is abstract and almost completely dependent on the experience and perception of teachers. "Vocal is the most difficult musical discipline, both for a teacher and a student, because both in the process of teaching and the learning process one has to spend a lot of time and a lot of effort and energy" Yamin (2017, p. 177). Using live recordings of singers, as teachers sometimes do, is not effective for students from China, since such "copying" of manner, voice and sound production can do more harm than help, especially at the initial stage. Such a method can complicate the process in the development and formation of the vocal apparatus, setting up the technological process, since everyone's physiology is different and requires an individual competent and sensitive approach (What suits me can harm him, and vice versa). It is important to be able to feel your body and be able to analyze its state, since the sound, its structure, etc. directly depend on this. Therefore, it is necessary to teach musical art by example, encouraging the student to reveal his individuality (but at the same time, when there are already basic skills, then comparison and knowledge of recordings of vocal works can help and be useful). A general overview of music education in China is as follows. Secondary professional music education in China lasts 3 years. Graduates receive a diploma of secondary vocational education, without qualifications. Complete higher musical education in China has 4 levels. The first level (2-3 years of study) gives the right to be called a specialist. After graduation, a diploma is issued. Studying at the second level lasts 4–5 years and entitles the graduate to receive a bachelor's degree. Bachelor's degree in music pedagogy lasts 4 years, and in conservatories, training lasts 5 years. Education at the third level - master's degree - is designed for 2-3 years. After successful completion of training, a master's degree is issued. The fourth level of study is doctoral studies, which lasts 3 years. After graduation, a Doctor of Science diploma is issued and a doctoral degree is awarded.

As part of the general programs of higher professional education in China, attention should be focused on the peculiarities of the vocal training of future music teachers, since this is an important factor in the design and implementation of study programs in Ukrainian universities, as well as in the development and implementation of a methodology for self-organization of the learning space of Chinese students. Thus, the specific features of the national culture of Chinese students, their mentality, as well as the level of general musical training and vocal training will be taken into account. The purpose of this chapter is to present a practical experience of intercultural communication between Ukrainian music teachers and students from China using ESP through writing. During the study, the following tasks were set: 1. To study the literature on ESP for musicians. 2. Conduct a survey of students from China using a questionnaire to determine the most convenient and intelligible form of communication. 3. Analyze the received data and recommend the use of the ESP letter for Ukrainian and Chinese musicians. When researching, it is important to take into account the socio-cultural characteristics of Chinese students, which affect their success in studying at higher musical institutions in Ukraine, revealing the peculiarities of the use of writing in the process of communication with teachers.

BACKGROUND

Literature Review

The problem of intercultural communication between music teachers and students is due to the difference between the Ukrainian and Chinese languages, so a fairly large amount of scientific sources from different fields of humanitarian knowledge are used to solve it. To solve it, the author of this chapter needed literature on the methodology of teaching a foreign language for Ukrainian specialists, as well as works on education in China, the typology of languages and innovative teaching aids, which became especially widespread during the covid-19 pandemic. So, for example, the article by Belyakov (2014) is devoted to teaching a foreign language in the system of training a Ukrainian performing musician. The work of Ilvitska, O. M. (2019) is devoted to the problems of teaching Chinese students in multinational groups of the preparatory department. In the study by Hu Zh. (2013), the definition of the socio-cultural characteristics of the adaptation of Chinese students to study at the universities of Ukraine is given. The problem of adapting to a new environment arises due to the difference in cultures. "A feature of Chinese culture is its belonging to the collectivist type. The main values of collectivist cultures are the imitation of traditions, obedience, and a sense of duty, which contribute to the preservation of the unity of the group, the interdependence of its members and harmonious relations between them" (Hu, Zh. 2013, p. 37). Students from China are characterized by the desire to help other people, prevent conflicts and maintain harmony. However, being in a foreign society, students feel disoriented.

Savchenko (2018) described the teaching of the Chinese language to Ukrainian students, and the work of Limin (2011) is devoted to the formation of the vocal culture of students of art institutes. The basic principles of teaching English to Chinese students are disclosed in the article by Peng, E., Woodrow, L., & Henze, J. (2021). The work of Gorchakov, O. A., & Qujia, Ch. (2011) on the management of the educational activities of Chinese students in Ukrainian universities is interesting. On the specifics of vocal education and the psychological characteristics of Chinese students, the works of Nagaitsev, N.I. (2021) and Si, D. (2015). Luhanska (2018), and Avramenko (2019) considered the features of intensive teaching of the Chinese language and analysis of mistakes that Ukrainian students make in the learning process. According to Nagaitsev, N.I. (2021), when presenting the content of his subject to foreign students, the teacher should not only use the methods and techniques worked out over the years and apply innovative technologies that intensify the learning process, but also take into account the regional socio-psychological characteristics of students. From the practice of teaching foreigners from around the world, it follows that students have striking features of the national character, such as restraint (bordering on coldness) and closeness. These traits have been formed over millennia and are the result of upbringing. For such students, the main subject is the teacher; Chinese students are closed to improvisation and competition in class. That is why in the Chinese audience it is important to form a sense of individual success in the group, which would allow students to overcome natural shyness and aloofness. For the positive development of intercultural communication, a large role is given to teaching foreign languages not only as a means of communication but also as a tool for learning the history and culture of other people and developing tolerance skills. "The tolerant type of speech behaviour is a linguoculturological category and involves the use of such communication tactics as compliance, compromise, cooperation, non-conflict, and politeness, which forms the basis for the formation of positive relationships" (Nagaitseva, 2021). The problems of adaptation and modern vectors of modern adaptation practices in teaching Chinese students in Ukraine are the subject of research by Zhongsi, X. (2016), Xiang, V. (2023, January), Gruciak (2015), and Bin, V. (2010). To select an innovative means of teaching music using writing in a foreign language of a specialized orientation, the author of this chapter processed the works of Morozov, Prokhorov & Makarov (2015), Antropov (2018), Tseryulnik & Zhai (2020), Nekrylov (2019), Tseryulnik (2020), Baranchuk & Klymenko (2019), Ignatovich, T. V. (2022).

Tetter, A. Yu., Kalina, Yu. V., Konovalchuk, A. A., & Abramchikova, L. V. (2021). Scientists write that one common group was created for students in the Chinese messenger WeChat, students sent assignments to the teacher in WeChat and in some cases using WeChat is more convenient than other programs. Ukraine is a European country, however, a significant part of its territory was influenced by Eastern culture for a long time, which left its mark on the mentality and peculiarities of the cultural life of Ukrainians. "This can become the basis for searching for common features of Ukrainian and Chinese cultures. However, Ukraine and China are two, in many respects, opposite types of cultures, they are two worldviews, two mentalities, and two languages, conditioned by different sociocultural values. Features are manifested not only in the diversity of spiritual culture but also in the way of life of our people as a whole. This is the attitude to the world, being, nature, the difference in the interpretation of values, the goals of human development, its purpose, understanding the meaning and orientation of human labour, the historical process" (Khu, Zh. 2013, p. 37). Hu Zh. (2013) notes that, unlike the Ukrainians, the Chinese manifest such national psychological qualities as strict discipline, high dependence on the group, specific unity based on a clear distribution of roles, a high degree of trust in the opinion of the group, as well as a special nature of compassion and empathy. As follows from the above study by Nagaitsev, N.I. (2021) it is the language that serves not only as a means of communication, but also as a tool for learning about the culture and developing professional communication skills, so the study of foreign languages becomes necessary, including from the standpoint of pedagogical and linguoculturological aspects. Language is a mirror of culture and reflects not only the real world and the conditions of human life, but also the public consciousness of the people, their mentality, national character, way of life, traditions, customs, morality, value system, attitude, and vision of the world. The language performs the function of a carrier and transmitter of the culture of the people, and transmits the treasures of national culture, traditions and customs. The Chinese language in our time is the most widely spoken in the world (Nagaitseva, 2021, p. 64). Thus, from the scientific sources discussed above, there follows an understanding of the relevance of studying written professional English by both music teachers and Chinese students, since this knowledge contributes to the quality teaching of musical subjects.

MAIN FOCUS OF THE CHAPTER

Typological Linguistic Characteristics in a Comparative Aspect

Music teachers in Ukraine usually take a professional foreign language course during their studies at music academies or institutes of music. They have developed the skills of oral speech, and reading professional texts, the vocabulary is sufficiently expanded, which implies the formation of the communicative competence of a specialist in the field of music. Training mainly takes place on various texts that contribute to the expansion of the horizons of students, the formation of writing, reading and speaking skills, and professional competence. Authentic textual material is usually selected on topics of musicological orientation and an active dictionary, as well as exercises for various activities. Creative exercises are developed for each text, which involves both works in a group under the guidance of a teacher, and independent work. The basics of business English are also captured, designed to form and activate business communication skills among musicians. Ukrainian musicians are also trained in the course of postgraduate professional education (postgraduate and assistant studies - internships). Thus, Ukrainian music teachers are competent enough to transfer their musical knowledge to students from other countries. If we talk in more detail about the acquisition of appropriate written skills, it is advisable to consider the features of the typology of the languages in question. In this case, we are talking about a logical chain of languages: Ukrainian (Russian) - English - Chinese. Kobyakova (2017) in her work "Introduction to the comparative typology of nominative and communicative units of the English, Russian and Ukrainian languages" notes that words are characterized by multidimensionality and the common thing for lexemes of compared languages is that they have a certain semantic volume, have structural features, form verbal series, is characterized by a certain motivation, form phraseological units, function in various styles, are used in terminological and non-terminological vocabulary (Kobyakova, 2017, p.6). "The unity of language and thinking, on the one hand, and their non-identity, on the other, is one of the main provisions of linguistics. The word, arising simultaneously with the concept, creates the possibility of generalization and abstraction. It fixes in our minds a generalized idea of reality. It is universal that thought is always looking for its linguistic expression, that each language has a special way of expressing thought, and that there is no exact coincidence of the semantic volume of lexical units of different languages. The objectively existing external world finds its reflection in consciousness - in the thinking of a person and its expression - in language. A thought expressed in one language can always be "reexpressed" using another language. Language is, first of all, a way of organizing information, while each language conveys the same information in different ways, associating it with specific markers" (Kobyakova, 2017, p. 7). Khakimzyanova, Khairullina, and Nesterenko write that the Chinese language is not single and universal. It includes many different dialects. Thus, ten dialect groups are distinguished in modern Chinese (Khakimzyanova et. al,, 2021, p. 66).

"Chinese dialects differ phonetically, lexically and grammatically, although the basics of their grammar and vocabulary are the same. Northern Chinese, also known as Mandarin, is the native language of about 70% of the Chinese and is the common written language for all Chinese" (Khakimzyanova et. al,, 2021, p. 66). English also has variants - British, American, Canadian, Australian, New Zealand, Irish, Indian and South African - based on the fact that there are differences in spelling, vocabulary and grammar depending on the country of the speaker. Belonging to two different language families, English and Chinese have many significant differences. This makes learning English a real challenge for native Chinese speakers. Differences between English and Chinese grammar (Khakimzyanova et. al,, 2021, p. 67), (Volkov and Guruleva, 2018) are described as follows: 1) Verbs and tenses. In English, most of the information is conveyed using auxiliaries and verbs: is/are/were, eat / eats / ate/ eaten, etc. Chinese remains largely unchanged and conveys meaning through word order, adverbs, or common understanding context. In Chinese, the concept of time is not used in different tenses and verb forms, as in English. It is for these reasons that Chinese speakers have problems with the complex system of English verbs. Here are some typical mistakes in verbs and tenses: What do you do? (instead of:

What are you doing?) (wrong timing) I will call you as soon as I will come home. (wrong timing) She has got fired last Saturday. (wrong timing) She good doctor. (missing linking verb) I wish I am tall. (indicative instead of subjunctive) There are no declensions or conjugations in Chinese, so the word order in a sentence plays a key role in the perception of meaning (Khakimzyanova et. al,, 2021, p. 67). Thus, in a Chinese sentence, the facts preceding the verb are first stated, and at the very end, the action itself is directly stated. 2) Modal verbs In English, shades of meaning, and emotional emphasis are usually expressed by modal verbs. For example, the increasing degree of politeness of the following instructions: Close the window, please. Could you close the window, please? Would you mind closing the window, please? Since Chinese modal verbs do not convey such a wide range of meanings, Chinese learners may not use English modal verbs sufficiently. This can lead to the fact that in the course of a conversation, in some cases, the speakers will seem peremptory or rude. However, there are also many modal verbs in the Chinese language, the only difficulty is that the modal verbs of Chinese are often very similar in meaning to each other and sometimes it is very difficult to distinguish and understand the fine line, shades of meaning. Often, the meaning of a modal verb can change depending on who the speech is addressed to and with what intonation the statement sounds (Khakimzyanova et. al,, 2021, p. 67), (Volkov and Guruleva, 2018).

Turning to the typology of the English, Russian and Ukrainian languages according to the research of Kobyakov (2017, p. 7), one can find that there is a certain number of lexically expressed concepts that differ in their scope. So, in English, there is a word denoting a line on which the names of the author, artist, photographer, etc. are placed. In Russian and Ukrainian, we find another name - signature, signature, which corresponds to English. by-line. The Russian word teenager denotes the age from 12 to 16 years old, and the English word teenager - the age from 13 to 19. The range of meanings assigned to specific words in languages is nationally specific. So, for example, the words of Russian. - man, Ukrainian - people in English correspond to such units as man, person, and woman. The indicated words in the Russian and Ukrainian languages are characterized by a wider range of meanings than the corresponding words in the English language (Kobyakova, 2017, p. 7). It is also noted that full-meaning words, proper names, substitute words and connective words are inherent in English, Russian and Ukrainian languages. The most used are full-valued and connective words. In English, linking words are more common than in Russian and Ukrainian, which is a marker of the analytical structure of the English language. “The study of word motivation not only lifts the veil over the mechanism of naming objects and phenomena of the surrounding world but also reveals word-formation tendencies that operate in a particular language system. In English, compared to Russian and Ukrainian languages, semantic derivation is more common - the formation of a new word by translating the stem into another paradigm of inflexion. Associated with this is the abundant phenomenon of polysemy in the English language. Polysemy - a property of most words in different languages - has received more development in English vocabulary than in Russian and Ukrainian, due to the abundance of monosyllabic root words in it (Kobyakova, 2017, p. 7). Khakimzyanova, Khairullina and Nesterenko, (2021, p. 67) say that there are no articles in Chinese, so difficulties with their correct use in English are very common. Word order in Chinese and English is very different. For example, in Chinese, questions are conveyed intonation - (there are 6 dons. One word can have different meanings depending on which don (in other words, in which of the 6 intonations) this word is pronounced. Look at the information on this topic in more detail if possible and (if you wish. These are the few things I remembered when I tried to learn Chinese); Subject and verb are not interchanged as they are in English. Nouns cannot be subject to change, as in English; and adverbs usually precede verbs, unlike in English where there are strict rules governing the position of such sentence elements.

English emphasizes sentence structure, while Chinese emphasizes meaning. It is very common in English to see one long sentence with long modifiers, including pronouns like “we”, “she”, and “they” in addition to “that” and “which” to avoid repetition. A sentence can be long and complex, but still clear enough to be understood. In Chinese, the situation is completely different. A long sentence in Chinese would be very complex and extremely difficult to understand. Therefore, only short sentences or long sentences divided into short phrases separated by commas can be found. It can be said that English sentences are usually long, while Chinese sentences are short (Khakimzyanova et. al,, 2021, p. 67), (Volkov and Guruleva, 2018). Nekrylova (2020) in her scientific work quite well explained the phonetic problems of Chinese students that Ukrainian teachers face. Working in a non-linguistic environment with Chinese students studying Russian, the researcher noticed that students find it difficult to understand spoken language and convey information. Although this chapter is aimed at demonstrating the importance of exclusively writing in a foreign language of a specialized orientation, it is important to note the reason for the author of the chapter to refer specifically to writing. This is because the difficulties in communicating with students are caused precisely by phonetics. Phonetic errors give rise to a certain language barrier and misunderstanding on the part of the listener. Violations in the pronunciation of stressed and unstressed vowels, the indistinguishability of hardness/softness and sonority/deafness of consonants are associated with the lack of formation of sound-letter correspondences among Chinese students and are also due to the specifics of the Ukrainian and Russian phonetic system, its pronunciation norms. Instead of Russian sounds, Chinese students pronounce either Chinese sounds close to them or replace them with sounds with a change in the place and method of formation (Nekrilova, 2020, p. 159; Nuzhna, A. Yu., 2018). “Errors of a phonological nature in the speech of Chinese students studying the Russian language distort the sound form and meaning, thereby complicating, and even violating, the communicative act. For example, afpobus (bus), beat (to be), mat (mother), cats (cats), luchka (pen). For the Chinese, knowledge of a foreign language is a ticket to life, an opportunity for prestigious work and career growth, so interest in learning foreign languages, including Ukrainian, is growing every year in China. Slavic languages present significant difficulties for Chinese students because the Russian and Chinese language systems differ at the phonetic, lexical and grammatical levels, as well as in the field of graphics" (Nekrilova, 2020, p.160).

Nuzhna (2018), and Nekrilova (2020) draw attention to the fact that in Chinese literary pronunciation, there are no voiced consonants, except for sonorants. Chinese consonants are characterized by the opposition of aspirated to non-aspirated. And for Russian consonants, as well as for Ukrainian ones, the opposition of deaf to voiced is typical. Therefore, Chinese students often replace Russian voiced [b], [d], [d] with Chinese non-aspirated voiceless plosives [r] [t] [k]. For example, [ba] - [ra], [du] - [tu] [go] - [kuo]. For example, [b] a [b] ka (folder), [d] body - body, y [d] point - duck, [b] e [g] in (Beijing). Difficulties for the Chinese are hard and soft consonants. For example, in the words I live, life, no Masha [w], [w] is pronounced as semi-soft. Many Chinese students insert a vowel [and] after a soft consonant in place of a consonant, or pronounce a syllable in place of a consonant letter - a combination of a consonant with a vowel [and]. For example, the word letter is pronounced: [letter] – letter; seven - [seven], pen - pen. A common mistake Chinese students make is mixing [s] and [sh] sounds. A typical mistake is mixing the sound [t] and the affricate [h']. The Russian sound [h '] is always pronounced softly, and the Chinese affricate [ts] is solid, unlike the Russian [h ']. For example, the final consonants of the words beat and whip. To distinguish between [t] and [h '], you can use the following exercises: at-ach, from-och, ut-uch, it-ich, mint-ball, copper-sword, sing-furnace, beat-scourge. Some students do not soften a consonant if it is followed by a soft sign. For example, the words family, role, here, skates, solid [m], [l], [s], [n] are pronounced in place of the letters m, l, x, n (Nekrilova, 2020, p. 161). Some students pronounce Russian [з] and [д] the same way, like voiceless Chinese [ts] or voiced [dz] (Nekrilova, 2020, p. 162), (Ignatovich, 2022, p. 156). For most students, the [r] sound is a big problem. There is no such sound in Chinese, so students often replace it with the sound [l]: nap[l]ime[l] (for example), k[l]asive (beautiful), int[l]esny (interesting). It is very difficult for Chinese students to pronounce the sound [r] in an isolated form. Words come to the rescue in which the sound [p] seems to rely on another sound: grass, firewood, drill, trolleybus, tractor, another. Such consonants before [p] provide support for the repulsion of the tongue and the correct positioning. In some provinces of China, not only the sound [r] but also the sound [l] is missing, so some students mix the consonants [n] and [l], for example,

June and juy [n] y (June and July), [n] eka[n]stvo (drug) (NEKRILOVA, 2020, p. 162; Nuzhna, 2018, Pisklova, 2015). In their study, Gorchakova, OA, & Quijia, Ch. (2011) write that among the difficulties in learning, Chinese students most of all noted insufficient knowledge of the language of instruction (Ukrainian). Most of the students noted that it is difficult to study. The main difficulties are caused by phonetics (for example, the letter "p", stress), grammar (the most difficult thing is case conjugation), as well as vocabulary (especially professional and scientific). Insufficient knowledge of the Ukrainian language by a Chinese student on the part of the teacher can be erroneously assessed as ignorance and misunderstanding of the subject being studied. The students also expressed the opinion that most often the difficulties in studying a particular subject are associated with the teacher, his skill in teaching and communication, and the desire to work with foreign students. Thus, the best option for communication between students and Ukrainian music teachers is English, in this case, written English, as discussed below.

Written Communication Using Interactive Communication Tools in the Process of Teaching Musical Skills

After the situation with the coronavirus COVID-19, which especially affected China, many students switched to a remote study format. The educational sector began to actively use Internet resources. Internet technologies allow the teacher to conduct training without being close to the student, and to use resources for distance learning of music, creating conditions for the full assimilation of the material. Distance learning is understood as a form of learning, and e-learning means learning tools that underlie the use of digital technology and allow you to gain knowledge with its help. "The use of mobile technologies based on mobile devices has confidently entered all spheres of our lives, including education. Mobile devices - smartphones, iPhones, etc. - begin to play the role of the "third screen" of information after television and computers. The massive use of mobile devices in advanced countries such as China makes them an excellent technical tool and learning tool for organizing and supporting the educational process" (Antropova, 2018, p. 218).

Figure 1. Advantages of the remote form (summarized and compiled by the author of this chapter)

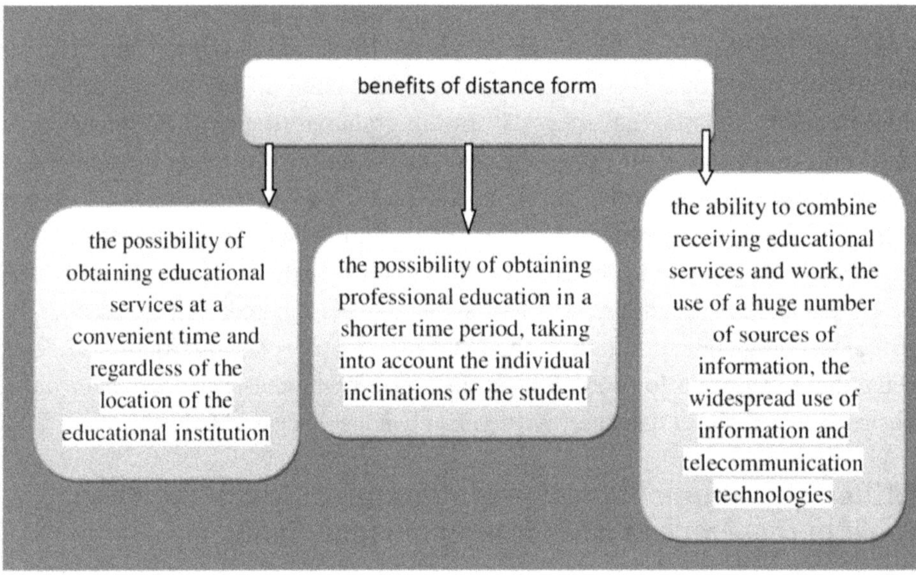

There are a large number of Internet resources, and programs that allow teachers to establish remote communication with students and organize the learning process in the most productive and sometimes non-standard way. The author of this study uses sending messages using WeChat. This mobile application is the subject of a scientific study by Antropov (2018). From it, it became known that WeChat is a mobile communication system for text messaging developed by the Chinese company Tencent, the first release of which was released in January 2011. "The application is available on the iPhone, as well as on Android, BlackBerry, WindowsPhone, Symbian and J2ME/S40 phones, there is also a web interface and a client for PC (Windows and OS X, which, however, require installation on a smartphone for authentication, however, this feature is available only to users of Android, iOS and Windows Phone). As of 2013, WeChat had 300 million registered users; of which 70 million are outside of China. The program was originally named Weixin but was renamed WeChat in April 2012 for the international market. To date, this program is number one in the world in terms of downloads and the number of users. This is a text and voice messenger with advanced features, including support for 18 languages, including Russian; the presence of elements of a social network; selection of funny stickers and games. To download the program, you can go to the official website www.wechat.com" (Antropova, 2018, p. 220). In the educational process of teaching music, WeChat uses: 1) in classroom work (class): as a visual learning tool, when a specific file (text, voice, music) or image (drawings, diagrams, tables, photos, videos) is transmitted through it to the group; as a control material (a file with a task for control, including tests) performed in the classroom; as a teaching material, if during the lesson you need to refer to any learning resource available on WeChat; as a result of student's project work, when a mobile presentation is performed either by a group or individually (Antropova, 2018, p. 222). 2) in extracurricular work: as a means of communication with members of a social network/group; obtaining visual and other information for project activities; checking and correct homework by types of speech activity, in particular, reading and listening (Antropova, 2018, p. 222). 3) for independent work: as a motivation to carry out project work and present its results using mobile presentations; as a source of information and other materials for the preparation of homework; to access RSL teaching resources to obtain educational and other materials necessary for successful learning (Antropova, 2018, p. 222).

Communication between the student and the teacher is carried out by writing messages in English. The letter in this case is the leading means of communication. In his work, Baranchuk (2020) writes that the multifunctionality of this application makes it an invaluable assistant in learning. "By sending and receiving text messages in the target language, students improve their writing and reading skills and learn to use the new language keyboard layout, which speeds up the process of immersion in a new language environment. The first words that students begin to exchange in a group chat are usually the words of greeting and farewell: "good morning", "good afternoon", "good evening", "hello", "hello", "goodbye", "bye", " see you tomorrow", "good night" (using an intermediary language or a translator, students can convey information about the areas, appropriateness and time of using each greeting and farewell: "hello / goodbye" with friends, classmates, "hello/goodbye" - with teachers, elders, strangers, etc.), as well as thanksgiving words: "thank you", "thank you very much" (Baranchuk, 2020, p. 398). Over time, students have the opportunity to practice using written English, asking questions and reading responses to messages from other group members, which also increases the effectiveness of musical learning material and makes the learning process more interesting and exciting. The teacher gets the opportunity to send lists of new tasks and even make this mailing a regular one. The ability to send voice messages contributes to the development of listening skills (when listening to an audio message) and speaking (when recording your audio messages). The teacher can use this function to record the correct musical or vocal sound. The presence of audio recordings and the ability to listen to them at any time an unlimited number of times increases the effectiveness of the acquired skills and helps to remember the correct musical sound and intonation design of vocal works. Students get the opportunity to listen, repeat and compare their pronunciation with the recording. The teacher can instantly listen to students' audio messages and give them feedback. Also, the teacher can combine the use of the functions of text and voice messages: for this, it is necessary to send a voice message and duplicate its content in a text message, thereby involving students in visual and auditory channels of information perception at the same time.

"Currently, most students do not like to write and consider this type of language activity boring and unnecessary. The task of the teacher is to create conditions under which writing becomes an interesting activity. Constant work with photos and videos allows students to collect a large album of impressions and memories, as well as educational materials on the language being studied" (Baranchuk, 2020, p. 399). The author of this study conducts vocal training classes based on the creative heritage of famous Ukrainian composers. The author of this chapter surveyed students with whom work continues in 2023. The survey included questions related to the educational repertoire, and ways of communicating in the WeChat program. Ancient arias of the 17th century, Italian songs, opera arias of the era of classicism, romances of the era of romanticism (Brahms, Strauss, Schubert, Schumann), folk art, romances to the music of Ukrainian composers: N. Lysenko, F. Nadenenko, Verikovsky and others. Using a letter in English, it was unanimously decided to communicate via WeChat. The survey results are shown in Figure 2.

Figure 2. The choice of musical material for the learning process by Chinese students as a result of a survey (conducted by the author of this chapter)

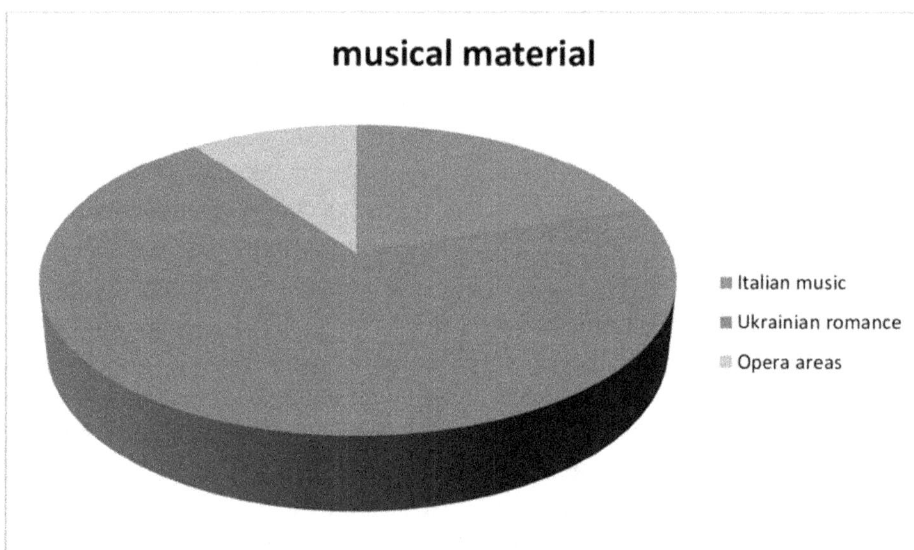

Summarizing the knowledge gained from the analysis of scientific literature regarding the peculiarities of the Ukrainian, Chinese and English languages, it was concluded that the simpler the recommendations are written in communication with Chinese students, the more productive the sound of musical works will be. Students write similar messages, clarifying certain tasks set by the teacher or accompanist for them. Below are some messages from the WeChat chat of the author of this chapter:

Dear students! This schedule is stable. It is advisable to follow it and be at your lesson on your own time! Enter it into your main study schedule so that there are no overlaps (trains, buses, meetings with friends and other misunderstandings). Please inform me about your absence from the lesson! If you can't be at the lesson, you can change the time with someone from the class, but ALWAYS tell me about this in advance. You see the entire timetable (whose lesson and when). We have a common group and you can independently agree with your colleagues about the changes. There must be discipline and stability in the class. Then we will have a good result in our work! Please note that the first column is your time in your time zone! Do not look at the second column! Good luck to everyone and see you in our classes!

1) Start this song.
2) You need to sing with your voice the sound that I press! Higher/lower!
3) Repeat again (this place/this work)
4) Start it over again.
5) Sing on / sing to the end.
6) You need to build a musical phrase with your voice. It's like telling a poem or an interesting story with an expression. It should be interesting to listen to you. This is the first part, the following... Tact is the second part. Both of these parts finish one big sentence in the first section.
7) The middle should be more mobile from the extreme parts./more calm and restrained from the extreme parts.
8) You sing with a flat sound, but you need to be more rounded.
9) Turn on my recording and sing this piece to it from the beginning to the end.
10) These commas mean that you need to take a breath in the diaphragm / in the stomach.
11) Put your breath in this place.
12) You can't take a breath in the middle of a word.
13) You need to sing these two tacts in one breath.
14) Don't exhale all the air at once. Take a breath, hold it and then start singing gradually blowing it out. Save your air.

A student of Hengshui University, who has experience working with Ukrainian teachers, expressed his opinion during this study that the language barrier is painful thing, because the translator only translates part of the meaning, and sometimes the sentence it translates is completely wrong. So it's important for student to have his own writing skills, to express the thoughts.

Thus, despite the ambiguous interpretation of the concept of the Internet language, the author of this chapter concludes that it is written English that is effective for communication between students from China who are mastering the art of music and Ukrainian teachers of musical disciplines. There is a kind of symbiosis between written English for specific purposes and Internet English. As Morozova, Prokhorova and Makarova (2015) write in their study: "This new Internet English is strikingly different from its prototype in that it is written only. However, being the language of live communication between people around the world, it cannot, by definition, be distinguished by rigour, extreme literacy, dryness and standard. Live communication between people is characterized by the desire for simplification, shortening and increased emotionality since the expression of feelings and emotions is a natural component of oral communication and is inseparable from human nature. Can we still say that Internet English is devoid of emotionality? A careful study of the correspondence between chats and social networks indicates the opposite. The lack of the ability to manipulate the rich phonetic arsenal of the English language opens up new opportunities for graphic expressive means, which essentially replaced the phonetic side of the language in the process of network communication" (Morozova et. al., 2015, p. 51).

CONCLUSION

As a result of the study, it was found that cooperation in music education between China and Ukraine has recently been quite high, despite the socio-political and economic upheavals. Chinese students need to have a higher education. Intercultural communication between Ukrainian teachers and Chinese students at the conservatory is carried out with the help of English, in particular, written. The problem of implementing joint programs of Chinese and Ukrainian higher education institutions was solved thanks to the developments that were developed during the Covid-19 pandemic. Knowledge of English of a professional musical orientation is necessary for Ukrainian music teachers and is provided during their studies at the conservatory, in the magistracy and in postgraduate studies. For Chinese students, English seems to be a prestigious field of humanitarian knowledge, which determines success in their future careers. Due to the presence of a large difference in the phonetic aspect, which is noted by many linguists. Thus, Chinese consonants are characterized by the opposition of aspirated to non-aspirated. Often Chinese students replace Russian voiced [b], [e], [d] with Chinese non-aspirated voiceless plosives [r] [t] [k]. Difficulties for the Chinese are hard and soft Ukrainian consonants. Students can insert a vowel

[and] after a soft consonant in place of a consonant, or pronounce a syllable in place of a consonant. Common mistakes made by Chinese students are mixing sounds [s] and [sh], and mixing sounds [t] and [h']. For most students, the [r] sound is a big problem. There is no such sound in Chinese, so students often replace it with the sound [l]. In some provinces of China, not only the sound [r] but also the sound [l] is missing, so some students mix the consonants [n] and [l]. Difficulties are also caused by grammar (the most difficult thing is case conjugation), and vocabulary (especially professional and scientific). There are a large number of Internet resources, and programs that allow teachers to establish remote communication with students and organize the learning process in the most productive and sometimes non-standard way. The author of this study uses a mobile communication system for text messaging developed by the Chinese company WeChat. In the educational process of teaching music, WeChat is used: in classroom work (in class), in extracurricular work, and for independent work. Communication between the student and the teacher is carried out by writing messages in English. Also, the teacher can combine the use of the functions of text and voice messages: for this, it is necessary to send a voice message and duplicate its content in a text message, thereby involving students in visual and auditory channels of information perception at the same time. It is also proved that in the process of distance communication between Chinese students and Ukrainian teachers of musical disciplines, it is written English that is the main and most acceptable form of professional communication. As a recommendation, we note that the simpler the messages are written, the better students acquire musical knowledge.

ACKNOWLEDGMENT

This research received no grant from any organization.

REFERENCES

Antropova, M. Y. (2018). Mobile technologies in the educational process (on the example of Chinese WeChat). *Cross Cultural Studies: Education and Science*, (3), 218–224.

Avramenko, B. V. (2019). Analysis of mistakes made by Ukrainian-speaking students in the process of learning Chinese phonetics. *Scientific Bulletin of the International Humanitarian University*, 38.

Baranchuk, O. (2020). Mobile technologies in teaching a foreign language. *Molody Vcheniy, 2*(78), 397-400.

Baranchuk, O. V., & Klimenko, M. P. (2018) Use of mobile technologies in teaching a foreign language. *Language teaching in higher education institutions at the modern stage. Interdisciplinary connections*, (33), 18-32.

Belyakova, N. S. (2014). The specifics of teaching a foreign language in the system of training a performing musician. *Education and upbringing: methods and practice*, (12), 149-153.

Bin, V. (2010). *The methodology for improving the Vikonavian technique of future music teachers in the processes of piano teaching* [Doctoral dissertation, Pivdennoukr. nat. ped. un-t im. K. D. Ushinsky] Pivdennoukr. nat. ped. un-t im. K. D. Ushinsky, Odessa. Ukraine. http://www.irbis-nbuv.gov.ua/cgi-bin/irbis_nbuv/cgiirbis_64.exe?C21COM=S&I21DBN=EC&P21DBN=UJRN&S21FMT=JwU_B&S21ALL=%28<.>U%3DIЩ315.42-712%20p%284YKP%293-3<.>%29&Z21ID=&S21SRW=AVHEAD&S21SRD=DOWN&S21STN=1&S21REF=10&S21CNR=20

Gorchakova, O. A. (2011). Management of the primary work of foreigners. (Chinese) students at the Higher Educational Institute of Ukraine. *Science and Education*, *7*, 82.

Grutsyak, V. I. (2015). Some features of the education of Chinese students at the preparatory department for foreign citizens. *Language teaching in higher education institutions at the modern stage. Interdisciplinary connections*, (27), 16-24.

Hu, J. (2013). Sociocultural features of adaptation of Chinese students to study at the universities of Ukraine. *Bulletin of Dnipropetrovsk University named after Alfred Nobel. Ser.: Pedagogy and psychology,* (1), 36-41.

Ignatovich, T. V. (2022). Teaching speaking in the absence of a language environment: difficulties and effective methods for overcoming them. In *Print technologies and media communications: materials of the 86th scientific and technical conference of faculty, researchers and graduate students, Minsk, January 31 - February 12.* Minsk: BSTU.

Ilvitska, O. M. (2019). Problems of training Chinese students in multinational training groups. *Ways to Science: Linguistic and Cultural Training of Foreign Students: Proceedings of the II International Scientific and Practical Conference,* 11-12.

Junsi, H. (2016). Vectors of modern practice of adaptation of Chinese students to study in Ukrainian universities. *Educational Expanse of Ukraine*, 7, 32-40.

Khakimzyanova, Yu.I., Khairullina, N.R., & Nesterenko, E.I. (2021). Comparative analysis of English and Chinese. *Eastern European Scientific Journal, 10*(74), 65-68.

Kobyakova, I. K., & Shvachko, S. A. (2017). *Introduction to comparative typology of nominative and communicative units of English, Russian and Ukrainian languages: teaching manual*. Sumy State University.

Lebiedieva, Z. D. (2023). The creativity of Fyodor Nadenenko in the space of the Ukrainian chamber and vocal repertoire. *Formation of science: state and prospects of development in today's conditions.*

Limin, V. (2011). Formation of vocal culture among students of art institutes on the basis of artistic traditions. *Youth and the Market,* (10), 159-162.

Luhanska, Z. A. (2018) *Features of intensive Chinese Learning in language courses.* Academic Press.

Morozova, E. N., Prokhorova, N. Yu., & Makarova, E. V. (2015). "Dry" language of internet communication "Barren" style of internet-communication. New in science: modern issues modern issues and trends, 50.

Nagaitseva, N. I. (2021). *Psychological features of studying students from China* [Doctoral thesis, LLC "Planeta-Print"]. Repository.

Nekrilova, O. L. (2020). From the practice of Russian movie phonetics for Chinese students: Problems and ability. *Mova*, (33), 159–163.

Nekrylova, E. L. (2019). The methodological potential of using the WeChat mobile application in teaching Russian to Chinese students. *Scientific Bulletin of the International Humanitarian University. Ser. Philology*, *40*(1), 72–75.

Nuzhna, A. Y. (2018). Problems of Chinese phonetics for University students. *Organizing Committee of the Conference*, 124.

Peng, E., Woodrow, L., & Henze, J. (2021). Basic principles for English movie Chinese students. *International Scientific and Practical Conference*, 28.

Pereloma, T. S. (2018). The use of information applications in the formation of phonetics competence of sinology students. *Collection of Scientific Works. Pedagogical Sciences*, *1*(85), 45–47.

Pisklova, Yu. A. (2015). Peculiarities of teaching phonetics of the Chinese language at the initial stages of learning. *Scientific Bulletin of Donbass*, (2).

Savchenko, A. S. (2018). Teaching Chinese language to students of the main foundations of enlightenment: the status of the problem. *Pedagogy and Psychology*, 33.

Si, D. (2015). The specifics of singing training of Chinese students in art institutes of pedagogical universities. *Scientific Journal of M. P. Drahomanov NPU. Series 14: Theory and Methods of Art Education, 18,* 210-213. http://nbuv.gov.ua/UJRN/Nchnpu_014_2015_18_51

Tatter, A. Y., Kalina, Y. V., Konovalchuk, A. A., & Abramchikova, L. V. (2021). Problems of implementing joint educational programs with Chinese Universities during the COVID-19 pandemic. Innovative projects and technologies in education, industry and transport, 399-408.

Tserulnik, A. Y. (2020). The functioning of the university in the context of distance learning during the COVID-19 pandemic. *Humanitarian Scientific Bulletin*, (5), 139–143.

Tserulnik, A. Y. (2020). The use of distance learning format for students in the educational process. *International Research Journal, 6*(96), 92-95.

Tserulnik, A. Y., & Zhai, C. (2020). Peculiarities of the use of We-CHAT and DINGTALK APPS in China for educational purposes and teachers. Basic and Applied Science: Status and Development Trends, 5-8.

Volkov, K. V., & Guruleva, T. L. (2018). Comparative analysis of the typological characteristics of the Chinese and Russian languages. *News of the Volgograd State Pedagogical University*, 8(131), 148–154.

Xiang, V. (2023, January). The problem of adaptation of student youth of the PRC to the conditions of study in institutions of higher education in Ukraine. In *The 1st International scientific and practical conference "Current issues of science and integrated technologies"*. International Science Group.

Yamin, I. (2017). Vocal and pedagogical enlightenment to China and its significance for the self-organization of the initial space of Chinese students. *Bulletin of Zhytomyr State University named after Ivan Franko, 1*(87), 173-177.

Zcin, N. (2009). *Methodical principles of vocal training of students from the People's Republic of China in the system of music-pedagogical education of Ukraine* [Doctoral thesis, National Pedagogical University named after M.P. Drahomanova] https://enpuir.npu.edu.ua/bitstream/handle/123456789/114/Jin_Nan.pdf?sequence=3

Zhao, W. (2014). *Methods of vocal and choral training of students in China with respect to Ukrainian singing traditions* [Doctoral dissertation, National Pedagogical University named after M.P. Drahomanova]. https://enpuir.npu.edu.ua/handle/123456789/6945

Compilation of References

Abramova, I. E., & Ananyina, A. V. (2010). Modernization of the language teaching system at the university. *Higher Education in Russia,* (8-9), 93-98.

Almusharraf, N., & Alotaibi, H. (2022). *An error-analysis study from an EFL writing context: Human and automated essay scoring approaches*. Technology, Knowledge, and Learning. doi:10.100710758-022-09592-z

Al-Sobhi, B. M. S. (2019). The nitty-gritty of language learners' errors—Contrastive Analysis, Error Analysis and Interlanguage. *International Journal of Education and Literacy Studies*, *7*(3), 49–60. doi:10.7575/aiac.ijels.v.7n.3p.49

Annett, C. H. (1982). An Editor's View: Ten Common Errors in Technical Writing. *Journal of Technical Writing and Communication*, *12*(3), 185–190. doi:10.2190/W558-WV3D-F1P6-HXMD

Antonova, T. V. (2017). Dictation in teaching a foreign language as a means of forming an integrated skill. *Bulletin of the Kostroma State University. Series: Pedagogy. Psychology. Sociokinetics*, *23*(3), 164–169.

Arıkan, A. & Yılmaz, A. F. (2019). Pre-Service English Language Teachers' Problematic Sounds. *International e-Journal of Educational Studies, 4*(7), 1-26. doi:10.31458/iejes.594715

Atmaca, Ç. (2016, April). Error analysis of Turkish EFL learners a case study. *GlobELT 2016 Conference*. 10.1016/j.sbspro.2016.10.007

Aufa, M., & Syarif, H. (2020). Error analysis on classroom language made by pre-service teachers of English education program UNP. *Journal of English Language Teaching*, *9*(1), 56. doi:10.24036/jelt.v9i1.108151

Bakhinskaya, I. A. (2021). Dictation as one of the ways for forming the discourse competence of technical students. *Linguistics and linguodidactics,* 14-17.

Baranova, N. A., & Ignatenko, E. V. (2016). Dictogloss as a means of activating the speech-cogitative activity of students in an English lesson. *Scientific and Methodological Electronic Journal "Concept", 12.* https://e-koncept.ru/2016/76137.htm

Baranova, N. A., & Ignatenko, E. V. (2016). Dictogloss as a means of activating the speech-cogitative activity of students in an English lesson. *Concept*, (S12), 1–5.

Bartholomae, D. (1984). The Study of Errors. *College Composition and Communication*, 31.

Bazanova, E. M. (2016). Scientific publication: write in English or translate? *Science Editor and Publisher, 1*(1-4), 17-24.

Belous, L. N. (2018). A modern approach to learning foreign words, or How to memorize new words in English. In Modern scientific research in the field of pedagogy and psychology (pp. 115-118). Academic Press.

Berninger, V. W., & Winn, W. (2006). Implications of advancements in brain research and technology for writing development, writing instruction, and educational evolution. In C. MacArthur, S. Graham, & J. Fitzgerald (Eds.), *Handbook of writing research* (pp. 96–114). Guilford Press.

Berry, A., Depaepe, F., & van Driel, J. (2016). Pedagogical content knowledge in teacher education. In J. Loughran & M. L. Hamilton (Eds.), *International Handbook of Teacher education* (pp. 347–386). Springer. doi:10.1007/978-981-10-0366-0_9

Bertkau, J. (1974). An Analysis of English Learner's Speech. *Language Learning*, *24*(2).

Bespalenko, O. N. (2012). Assessment of restoration dynamics of burnt forest area vegetation using Landsat satellite data. In Modern problems of land management, cadastre and land monitoring: textbook. VGLTA.

Bezruchko, E. N. (2011). English for architects: Textbook for universities in the direction of "Architecture" (3rd ed.). Academic Press.

Blinov, M. A. (2017). *On the criteria of objectivity and subjectivity in assessing written communication skills in English of students of AF SPbGUP*. Academic Press.

Britton, W. E. (1965). What Is Technical Writing? *College Composition and Communication*, *16*(2), 113. doi:10.2307/354886

Brooks, N. H. (1964). *Language and language learning; theory and practice*. Harcourt, Brace and World.

Brown, D. H., & Lee, H. (2001). *Teaching by principles: An interactive approach to language pedagogy*. Pearson Education ESL.

Brown, H. D. (2000). *Principals of language learning and teaching*. Longman.

Brown, H. D. (2001). *Teaching by Principle: An Interactive Approach to Language Pedagogy*. Longman.

Brown, H. D. (2014). *Principles of language learning and teaching: A course in Second language acquisition* (6th ed.). Pearson Education.

Budagyan, I. G., Vishnevetskaya, N. A., & Romanova, O. N. (2015). Transfer of features of the translation of texts of architectural specialties. *Business. Education. Law*, (4), 351–354.

Burt, M., & Kiparsky, C. (1972). *The Gooficon: A repair manual for English.* Newbury House Publishers.

Castillo Rodríguez, C., & López Pérez, S. (2019). To be or not to be...? analysis of misuses with be in a learner corpus in an online environment. *Revista De Filología De La Universidad De La Laguna*, (39), 153–170. doi:10.25145/j.refiull.2019.39.06

Castillo-Montoya, M. (2016). Preparing for Interview Research: The Interview Protocol Refinement Framework. *Qualitative Report*. Advance online publication. doi:10.46743/2160-3715/2016.2337

Chan, A. Y. (2010). Toward a taxonomy of written errors: Investigation into the written errors of Hong Kong Cantonese ESL learners. *TESOL Quarterly*, *44*(2), 295–319. doi:10.5054/tq.2010.219941

Chudley, R., & Greeno, R. (2006). *Building Construction Handbook*. Elsevier. doi:10.4324/9780080460864

Connolly, A. J. (2020). Adjective-noun order: An error analysis of Colombian learners of English. *GIST – Education and Learning Research Journal*, *20*, 231–250. doi:10.26817/16925777.708

Corder, J. P. (1981). *Error Analysis and Interlanguage*. Oxford University Press.

Corder, S. P. (1967). The Significance of Learner's Errors. *IRAL -. International Review of Applied Linguistics in Language Teaching*, *5*(1–4). Advance online publication. doi:10.1515/iral.1967.5.1-4.161

Corder, S. P. (1967). The significance of learners' errors. *International Journal of Applied Linguistics*, *5*(4), 161–170.

Corder, S. P. (1973). *Introducing applied linguistics*. Penguin.

Corder, S. P. (1975). Error Analysis, Interlanguage and Second Language Acquisition. *Language Teaching & Linguistics: Abstracts*, *8*(4), 201–218. doi:10.1017/S0261444800002822

Creswell, J. W. (2002). *Research Design: Qualitative, quantitative, and mixed methods approaches* (2nd ed.). SAGE Publications.

Creswell, J. W. (2012). *Educational research: Planning, conducting, and evaluating quantitative and qualitative research* (4th ed.). Pearson.

Creswell, J. W. (2013). *Qualitative inquiry and research design: Choosing among five approaches*. Sage Publications.

Crosthwaite, P. (2018). Does EAP writing instruction reduce L2 errors? evidence from a longitudinal corpus of L2 EAP essays and reports. *International Review of Applied Linguistics in Language Teaching*, *56*(3), 315–343. doi:10.1515/iral-2016-0129

CV and Cover Letter Advice. (2023). https://www.jobs.ac.uk/careers-advice/cv-templates

Dewi, R. C., & Fitria Hardini Rangkuti, S. (2021). Grammatical Errors on EFL Students' Conversation Practice: Surface Strategy Taxonomy. *Journal Of Education And Teaching Learning*, *3*(3), 10–17.

Dobrin, D. N. (1989). *Writing and Technique*. Natl Council of Teachers of English.

Dulay, H., & Burt, M. (1974). Errors and Strategies in Child Second Language Acquisition. *TESOL Quarterly*, *8*(2), 8. doi:10.2307/3585536

Dulay, H., & Burt, M. (1974). Natural sequences in child second language acquisition. *Language Learning*, *24*(1), 37–53. doi:10.1111/j.1467-1770.1974.tb00234.x

Dulay, H., Burt, M., & Krashen, S. (1982). *Language Two*. Oxford University Press.

Egorova, E. V., Krasheninnikova, E. I., Krasheninnikova, N. A., & Kuznetsova, O. Yu. (2018). Dictation in English (Total Dictation): Writing experience and results. Lingua Academica: Actual problems of linguistics and linguodidactics, 126-130.

Egoshina, N. G. (2022). Creating a cultural and educational space in the context of foreign language education. *Bulletin of the Perm National Research Polytechnic University. Problems of Linguistics and Pedagogy,* (3), 181-191.

Ellis, R. (1985). *Understanding Second Language Acquisition*. Progress Press.

Ellis, R. (1997). *Second language acquisition*. Oxford University Press.

Ellis, R. (2003). *The Study of Second Language Acquisition*. Oxford University Press.

Ellis, R. (2008). *The study of Second language acquisition*. Oxford University Press.

Ellis, R., & Barkhuizen, G. (2014). *Analysing learner language*. Oxford University Press.

El-Mahdy, F. M. (2023). An Error analysis of the grammatical errors of Egyptian EFL learners and a suggested program for enhancing their grammatical competence. *Transcultural Journal of Humanities and Social Sciences, 4*(1).

Emmel, N. (2013). *Sampling and choosing cases in qualitative research: A realist approach* (1st ed.). SAGE Publications Ltd. doi:10.4135/9781473913882

Emmit, S., & Gorse, A. Ch. (2005). *Barry's introduction to construction of buildings*. Blackwell Publishing LTD.

Erdoğan, V. (2005). Contribution of error analysis to foreign language teaching. *Mersin Üniversitesi Eğitim Fakültesi Dergisi, 1*(2).

Eunson, B. (2011). Communicating in the 21st Century 3E iStudy Version 1 (3rd ed.). Wiley.

Falaleeva, A. S. (2021). "Running dictation" as an effective method of conducting dictations in English lessons. *Modern Problems of Linguodidactics*, 126-129.

Farrell, T. (2002). Lesson Planning. In J. Richards & W. Renandya (Eds.), Methodology in Language Teaching: An Anthology of Current Practice (pp. 30-39). Cambridge University Press. doi:10.1017/CBO9780511667190.006

Federal State Educational Standard of Higher Education in the areas of training "41.03.05. International relations", "41.03.01. Foreign Regional Studies". (2023). https://eng.kpfu.ru/bachelor-specialist-programs/arts-and-humanities-bachelor-programs/

Flowerdew, J. (2016). English for Specific Academic Purposes (ESAP) Writing. *Writing & Pedagogy*, *8*(1), 1–4. doi:10.1558/wap.v8i1.30077

Frolova, N. A., & Aleshchanova, I. V. (2007). Training in abstracting and annotating professional literature in a foreign language in the complex training of a specialist. *Modern Problems of Science and Education*, (5), 11-11.

Fumero, K., & Wood, C. (2021). Verb errors in 5th-grade English learners' written responses: Relation to writing quality. *Languages (Basel, Switzerland)*, *6*(2), 71. doi:10.3390/languages6020071

Galbraith, D. W. (2012). *Writing about what we know: Generating ideas in writing*. SAGE Publications Ltd. doi:10.4135/9780857021069.n4

Gass, S. M., Behney, J., & Plonsky, L. (2020). *Second language acquisition: An introductory course*. Routledge. doi:10.4324/9781315181752

George, V. (1972). *Common Errors in Language Learning: Insights from English*: Rowley: Newbury House.

Gill, A. E., & Janjua, F. (2020). Genre pedagogy and ELLs' writing skills: A theme analysis. *English Language Teaching*, *13*(8), 141. doi:10.5539/elt.v13n8p141

Golovina, N. V. (2012). The significance of the interdisciplinary aspect in the formation of translation activities among students of specialized classes of secondary school. *Bulletin of the Tambov University. Series. Humanities*, (9), 175–179.

Google translate. (2023). https://translate.google.com/?hl=ru

Graves, S. (2010). Mentoring pre-service teachers: A case study. *Australasian Journal of Early Childhood*, *35*(4), 14–20. doi:10.1177/183693911003500403

Hamburger, H. (1981). *The Function of the Predicate in the Fables of Krylov: A Case Grammatical Study*. Rodopi.

Hanania, E. A., & Gradman, H. L. (1977). Acquisition of English Structure. *Language Learning*, 27.

Holcomb, Z. C., & Cox, K. S. (2017). *Interpreting basic statistics: A workbook based on excerpts from journal articles*. Taylor & Francis. doi:10.4324/9781315225647

How to write a CV. (2023). https://www.prospects.ac.uk/careers-advice/cvs-and-cover-letters/how-to-write-a-cv

Hunt-Gómez, C. I., & Navarro-Pablo, M. (2020). Analysis of pre-service foreign language teachers' incorrect articulations: Frequency, influence on communication, and a specific corrective strategy. *Problems of Education in the 21st Century, 78*(6), 933–947. doi:10.33225/pec/20.78.933

Hyland, K. (2003). *Second Language Writing* (1st ed.). Cambridge University Press. doi:10.1017/CBO9780511667251

Hyland, K., & Hamp-Lyons, L. (2002). EAP: Issues and directions. *Journal of English for Academic Purposes, 1*(1), 1–12. doi:10.1016/S1475-1585(02)00002-4

Islamov, R. S. (2020). Artificial intelligence in information and communication technologies and its impact on foreign language teaching in higher education. *Philological Sciences. Questions of Theory and Practice, 13*(12), 300–305.

James, C. (1998). *Errors in Language Learning and Use*. Longman.

James, C. (1998). *Errors in Language Learning and Use: Exploring Error Analysis (Applied Linguistics and Language Study)* (1st ed.). Routledge.

James, C. (2013). *Errors in language learning and use: Exploring error analysis*. Routledge. doi:10.4324/9781315842912

Jin, F. (2009). Third language acquisition of Norwegian objects: Interlanguage transfer or L1 influence. In Y. I. Leung (Ed.), *Third language acquisition and universal grammar* (pp. 144–161). Multilingual Matters. doi:10.21832/9781847691323-010

Kalashnikova, A. A., & Akai, O. M. (2016). Methodology of translation and teaching translation from a foreign language in a non-linguistic university. *Modern High Technologies, 10*(2), 337-341.

Kapterev, A. A. (2012). *Presentation skill. How to create presentations that can change the world from English*. Mann, Ivanov and Ferber, Eksmo.

Karney, B. W., & Filion, Y. (2000). Technical writing: Purpose and principles. *Journal of Hydraulic Engineering, 126*(4), 230–231. doi:10.1061/(ASCE)0733-9429(2000)126:4(230)

Keshavarz, M. H. (1994). *Contrastive analysis and error analysis*. Rahnama Publications.

Khalilova, L.A. (2007). *Foreign language in a non-linguistic University: what, why and how to teach.* GEOTAR-Media.

Kokoreva, A. A. (2013). Corpus of parallel texts in foreign language teaching. *Bulletin of the Tambov University. Series. Humanities*, 2(118), 57–62.

Kolesnichenko, A. N. (2021). The use of digital technologies in teaching the vocabulary of a foreign language in higher education. *Pedagogy. Questions of Theory and Practice, 6*(2), 298–303.

Konovalova, E. N., Marsheva, F. M., & Mubarakshina, F. D. (2014). *Professional English in use: Architecture. Part 1.* Kazan.

Kornienko, O. P. (2008). Terminological specificity of English texts of the sublanguage of architecture and construction. *Proceedings of the Russian State Pedagogical University*, 63-1.

Korotkikh, Ya. I., Mazalova, V. Yu., & Malinitskaya, E. P. (2019). Computer translation and its use in the educational process in a military university when studying a foreign language. Innovative educational space: Theory and practice of teaching foreign languages and Russian as a foreign language in higher education, 109-121.

Krashen, S. (1975). An error analysis of an advanced ESL learner: The importance of the monitor. *Working Papers on Bilingualism, 7*, 125–129.

Kroll, B. (1990). *Second language writing: research insights for the classroom.* Cambridge University Press. doi:10.1017/CBO9781139524551

Kulac, O., Babaoglu, C., & Akman, E. (2022). Public Affairs. *Education + Training*, 21.

Kurt, A. (2021). *Self-Determination and its relationship to conscientiousness, career motivation, gender, and the year of study in pre-service English language teachers* (Publication No: 677265) [Doctoral Dissertation, Onsekiz Mart University]. YÖK Ulusal Tez Merkezi. https://tez.yok.gov.tr/UlusalTezMerkezi

Kuznetsova, N. G., Zaitseva, I. E., & Stepicheva, O. N. (2016). The Professional Language of Architects: A Systematic Approach to the Dictionary (Based on English, German and French). *Philological Sciences. Questions of Theory and Practice, 6*(60), 98-107.

Kuznetsov, V. A. (2006). Scientific adviser: Savinov AN. *Food Technology*, (2-3), 70.

Lado, R. (1957). *Linguistics Across Cultures: Applied linguistics for language teachers*. University of Michigan Press.

Larissa, B., & Valeria, C. (2019). Technical Writer in the Framework of Modern Natural Language Processing Tasks. *Journal of Siberian Federal University. Humanities & Social Sciences, 12*(1), 20–31. doi:10.17516/1997-1370-0377

Lebedeva N. A. (2017). Information aspects in the context of innovative means of teaching a foreign language. *All-Ukrainian science and practice internet conference "Modern management: problems and development prospects on May 19, 2017"*, 235-238.

Lebedeva, N. (2019). Information and communication technologies for teaching a foreign language regarding the modernization of public administration. *The latest information and communication technologies in the modernization of public administration: foreign and domestic experience: materials of II scientific practice seminar.* Dnipropetrovsk DRIDU NADU.

Lebedeva, N. A. (2018). Teaching English in the field of knowledge "Public management and administration" in an agrarian higher educational institution. In *European dimension of reforming public administration in Ukraine: materials of the International Scientific and Practical Conference.* Kyiv Publishing House "Personal".

Lebedeva, N. A. (2022). Teaching of Foreign Language in the Educational Process of Public Administration Personnel in Ukraine. In O. Kulaç, C. Babaoğlu, & E. Akman (Eds.), *Public Affairs Education and Training in the 21st Century* (pp. 294–310). IGI Global. doi:10.4018/978-1-7998-8243-5.ch019

Lee, J. (2020). An error analysis of L2 writing: How does task complexity affect L2 errors? *The Journal of AsiaTEFL*, *17*(3), 954–965. doi:10.18823/asiatefl.2020.17.3.13.954

Lelakova, E., & Toman, M. (2023). A corpus based study of commas use in EFL written performance. *International Journal of Instruction*, *16*(1), 1–20. doi:10.29333/iji.2023.1611a

Leong, L. M., & Ahmadi, S. M. (2017). An analysis of factors influencing learners' English speaking skill. *International Journal of Research in English Education*, *2*(1), 34–41. doi:10.18869/acadpub.ijree.2.1.34

Levina, L. V., & Levin, A. I. (2019). Ensuring continuity in the process of teaching the translation of authentic texts from a foreign language into Russian in a non-linguistic university. In *Foreign Language Teaching* (pp. 224–233). Modernity and Perspectives.

Lightbown, P. M., & Spada, N. (2006). *How languages are learned.* Oxford University Press.

Likhacheva, O. N., & Bedenko, D. E. (2020). Teaching a foreign language in modern conditions in the context of a technical university. *Electronic network polythematic journal "Scientific works of KubGTU,* (5), 70-76.

Lipatova, V. V., & Litvinov, A. V. (2011). The concepts of equivalence and adequacy in teaching translation in higher education at the present stage. *Russian Journal of Linguistics*, (4), 105–113.

Lococo, V. (1976). A Comparison of Three Methods for the Collection of L2 data: Free Composition, Translation, and Picture Description. *Working Papers on Bilingualism*, 8.

Longcope, P. (2009). Differences between the EFL and the ESL language learning contexts. *Studies in Language and Culture*, *30*(2), 203–320.

Manchon Ruiz, R.M., Murphy, L. & Roca de Larios, J. (2007). Lexical retrieval processes and strategies in second language writing: a synthesis of empirical research. *International Journal of English Studies*, 7.

Manchon Ruiz, R. M., Murphy, L., & Roca de Larios, J. (2007). Lexical retrieval processes and strategies in second language writing: A synthesis of empirical research. *International Journal of English Studies*, *7*, 149–174.

McDowell, L., & Liardet, C. (2020). Towards specialized language support. An elaborated framework for error analysis. *English for Specific Purposes*, *57*, 16–28. doi:10.1016/j.esp.2019.09.001

McKinley, J. (2015). Critical Argument and Writer Identity: Social Constructivism as a Theoretical Framework for EFL Academic Writing. *Critical Inquiry in Language Studies*, *12*(3), 184–207. doi:10.1080/15427587.2015.1060558

Melentieva, O. Y. (2009). The development of spelling vigilance at the initial stage of learning English. *Municipality: Innovation and experiment,* (2), 72-75.

Molodykh, O. V. (2021). Teaching spelling as a means of mastering written speech in English lessons. *Journal of Scientific Conferences, 2*(1), 81-82.

Montrul, S. (2011). Morphological errors in Spanish second language learners and heritage speakers. *Studies in Second Language Acquisition, 33*(2), 163–192. doi:10.1017/S0272263110000720

Morgan, H. (2022). Conducting a Qualitative Document Analysis. *Qualitative Report, 27*(1), 64–77. doi:10.46743/2160-3715/2022.5044

Morse, J. M. (2000). Determining sample size. *Qualitative Health Research, 10*(1), 3–5. doi:10.1177/104973200129118183

Murphy, L. & Roca de Larios, J. (2001). Some steps towards a socio-cognitive interpretation of second language composition processes. *International Journal of English Studies,* 1.

Murphy, L., & Roca de Larios, J. (2001). Some steps towards a socio-cognitive interpretation of second language composition processes. *International Journal of English Studies, 1*, 25–45.

Murzin, A. D. (2013). *Real estate: Economics, valuation and development.* Phoenix.

Musorina, O. A., & Sorokina, S. G. (2019). *Organization of educational work at the university at the level of the department: concept, content, methods for evaluating effectiveness. In Education: experience and development prospects. Chuvash Republican Institute of Education.* Publishing House LLC Publishing House Sreda.

Mustafina, F. Sh. (2014). Dictation as a means of improving graphic and spelling skills in teaching foreign languages. *English language. Everything for the teacher!, 32*(8), 6-7.

Nagiyatova, L. B. (2018). Learning to write: Dictation in an English lesson in a non-linguistic university. *Alley of Science, 4*(11), 185–188.

Najlah, N., & Fatimah, S. (2020). Common Grammatical Errors in Oral Communication Made by Indonesian EFL Learners. *Journal of English Language Teaching, 9*(4), 740–749. https://doi.org/https://doi.org/10.24036/jelt.v9i4.110341

Nikolenko, O., Rebenko, M., & Doronina, N. (2021). Academic literacy: The analysis of first-year Ukrainian University Students' errors in essay writing. *Revista Romaneasca Pentru Educatie Multidimensionala, 13*(1). doi:10.18662/rrem/13.1Sup1/396

Nunan, D. (1991). *Language teaching methodology: A textbook for teachers.* Longman.

Nurfaidah, S. (2018). Vygotsky's Legacy on Teaching and Learning Writing as Social Process. *Langkawi: Journal of the Association for Arabic and English.* doi:10.31332/lkw.v4i2.1038

Nzama, M. V. (2010). Error Analysis: A Study of Errors Committed by Isizulu Speaking Learners of English in Selected School. *International Journal of Humanities and Social Science, 8*, 116–125.

Osipova, V. S., & Putistina, O. V. (2021). The use of dictations in English lessons as a method of activation of students' speech thinking activity. *Collection of the Regional Scientific-Practical Conference-Student Scientific School of the MASU branch in Apatity*, 97-102.

Pahomkina, M. E. (2013). The role of translation in teaching a foreign language in the magistracy. *Proceedings of the Southern Federal University. Engineering Sciences*, (10), 99-103.

Pasko, O. A., & Kovyazin, V. F. (2017). *Research work of a master student: teaching aid.* TPU.

Pasko, O., Kovyazin, V., Zakharchenko, A., & Lebedeva, N. (2020). Spatial organization of forestfund. *IOP Conference Series. Earth and Environmental Science*, *574*(1), 012061. doi:10.1088/1755-1315/574/1/012061

Pasko, O., Staurskaya, N., Zakharchenko, A., Zharnikov, V., & Larionov, Y. (2022). *Bio-Farming as the Basis of Environmentally-Sustainable Arable Farming at the Time of Global Warming.* Research Anthology on Strategies for Achieving Agricultural Sustainability. doi:10.4018/978-1-6684-5352-0.ch031

Pasko, O., Zakharchenko, A., & Kovyazin, V. (2021). *Analysis of land management of the forest fund.* Bulletin of the Tomsk Polytechnic University. *Geo Assets Engineering.*, *332*(2), 127–138.

Peng, S. (2019). A Study of the Differences between EFL and ESL for English classroom teaching in China. *IRA International Journal of Education and Multidisciplinary Studies*, *15*(1), 32–35. doi:10.21013/jems.v15.n1.p4

Peter, J., & Singaravelu, G. (2021). Problems in writing in English among high school learners. *AEGAEUM Journal*, *8*(9). 1502-1515. https://www.researchgate.net/publication/351048985_Problems_in_Writing_in_English_among_High_School_Learners

Petrosyan, I. V. (2019). The work program of the discipline "Practical course of the first foreign language" (for 1st year undergraduate students). *Collection of Programs*, 366-410.

Pilyugina, O. P. (2005). Scientifi c processing of professional texts//Technologies of teaching foreign languages in non-linguistic Universities. In *Collection of scientifi c articles of interuniversity scientifi c and technical conference.* Ulyanovsk: USTU.

Polat, E. S., & Bukharkina, M. Yu. (2007). Modern pedagogical and information technologies in the education system. Ed. Center "Academy".

Powell, C. K., & Kalina, J. C. (2009). Cognitive and social constructivism: Developing tools for an effective classroom. *Education*, *130*, 241–250. doi:10.4236/ojapps.2022.121011

Priya, M. R., & Divya, K. V. (2016). Writing skills for technical writers. *International Journal of Engineering Technology, Management and Applied Sciences*, *4*(2). https://www.researchgate.net/publication/304150238_Writing_Skills_For_Technical_Writers

Proskuryakova, N. A. (2013). Means of developing the abilities of students-future architects. *News of the Samara Scientific Center of the Russian Academy of Sciences*, *15*(2-4), 914–916.

Purgason, K. B. (2014). Lesson planning in second/foreign language teaching. In M. Celce-Murcia, D. M. Brinton, & M. A. Snow (Eds.), Teaching English as a second or foreign language (pp. 362–379). National Geographic Learning, Heinle Cengage Learning.

Ramón, N. (2006). Multiple Modification in English and Spanish. *International Journal of Corpus Linguistics*, *11*(4), 463–495. doi:10.1075/ijcl.11.4.06ram

Rana, L. B. (2018). *L1 Use in L2 Writing: A Case of Multilingual Undergraduates in a Basic Writing Course* [Master's thesis, Minnesota State University, Mankato]. Cornerstone: A Collection of Scholarly and Creative Works for Minnesota State University, Mankato. https://cornerstone.lib.mnsu.edu/etds/786/

Rao, V. C. S. (2020). Academic language proficiency: English for Academic Purposes (EAP). *Journal for Research Scholars and Professionals of English Language Teaching*, *21*(4). https://files.eric.ed.gov/fulltext/ED611477.pdf

Rehman, I., Silpachai, A., Levis, J., Zhao, G., & Gutierrez-Osuna, R. (2020). The English pronunciation of Arabic speakers: A data-driven approach to segmental error identification. *Language Teaching Research*, *26*(6), 1055–1081. doi:10.1177/1362168820931888

Reid, J. M. (2001). The Cambridge guide to teaching English to speakers of other languages. In R. Cater & D. Nunan (Eds.), *Writing* (pp. 23–33). Cambridge University Press.

Richards, J.C. (1971). A Non-Contrastive Approach to Error Analysis. *Journal of ELT*, 25.

Richards, J. C. (1973). *Error Analysis: Perspectives on Second Language Acquisition*. Longman.

Richards, J. C. (1974). *Error Analysis: Perspective on Second Language Acquisition*. Longman.

Richards, J. C., & Renandya, W. A. (2002). *Methodology in language teaching: An anthology of current practice*. Cambridge University Press., doi:10.1017/CBO9780511667190

Richards, J. C., & Schmidt, R. (2010). *Longman Dictionary of Language Teaching and Applied Linguistics*. Longman.

Roca de Larios, J. (1996). Linearization strategies in ELF writing. *Lenguaje y textos*, *8*, 191-208.

Roca de Larios, J. (1996). Linearization strategies in ELF writing. *Lenguaje y textos*, 8.

Roca de Larios, J., Manchon Ruiz, R. M., & Murphy, L. (2007). Componentes básicos y evolutivos del proceso de formulación en la escritura de textos en lengua materna y lengua extranjera. *Revista Española de Lingüística Aplicada*, 20.

Romanova, O. N., Astafurova, T. N., Vishnevetskaya, N. A., & Surkova, E. V. (2022). Innovative technologies in teaching a foreign language to architect technologies in teaching a foreign language to architect students. *News of the Volgograd State Pedagogical University*, *5*(168), 56–60.

Rustipa, K. (2011). Contrastive analysis, error analysis, interlanguage and the implication to language teaching. *Ragam Jurnal Pengembangan Humaniora*, *11*(1), 16–22.

Sağlamel, H., & Aydoğdu, Z. M. (2021). The Academic writing needs of students: A case study on stakeholder perspectives. Acuity. *Journal of English Language Pedagogy, Literature and Culture*, *7*(1), 37–50. doi:10.35974/acuity.v7i2.2541

Saito, K., & Hanzawa, K. (2018). The role of input in second language oral ability development in foreign language classrooms: A longitudinal study. *Language Teaching Research*, *22*(4), 398–417. doi:10.1177/1362168816679030

Sartika, D. (2022). A Correlation between Students' Self-Esteem and Speaking Skill. *JETLe*, *3*(2), 1–9. doi:10.18860/jetle.v3i2.15610

Sarwat, S., Ullah, N., Anjum, H. M. S., & Bhuttah, T. M. (2021). Problems and factors affecting students English writing skills at elementary level. *Elementary Education Online*, *20*(5), 3079–3086. doi:10.17051/ilkonline.2021.05.332

Saville-Troike, M. (2012). *Introducing Second Language Acquisition*. Cambridge University Press. doi:10.1017/CBO9780511888830

Schreiber, L. M., & Valle, B. E. (2013, May 29). Social Constructivist Teaching Strategies in the Small Group Classroom. *Small Group Research*, *44*(4), 395–411. doi:10.1177/1046496413488422

Selinker, L. (1972). Interlanguage. *International Review of Applied Linguistics in Language Teaching*, *10*(1–4). Advance online publication. doi:10.1515/iral.1972.10.1-4.209

Selinker, L. (1992). *Rediscovering Interlanguage*. Longman.

Senko, Yu. V., & Frolovskaya, M. N. (2007). *Pedagogy of understanding*. Drofa.

Seow, A. (2002). The Writing Process and Process Writing. In J. C. Richards & W. A. Renandya (Eds.), *Methodology in language teaching* (pp. 315–327). Cambridge University Press. doi:10.1017/CBO9780511667190.044

Shalnev, N. A. (2014). Conducting dictations in foreign languages as a form of organization of classroom and extracurricular activities of students. *Languages of the peoples of the world and the RF,* 186.

Shcherbakova, I. V., & Timashova, M. V. (2019). The role of professionally oriented education in the development of dialogic and monologue speech in a foreign language in a non-linguistic university. *Baltic Humanitarian Journal*, *8*(3), 205–208.

Shevchenko, M. A., & Mitchell, P. D. (2013). Training of military translators in a civilian university (experience of the National Research Tomsk State University). *Language and Culture*, 1(21), 125–131.

Shishkina, T. G. (2020). The work program of the discipline "Practical course of the 1st foreign language (English) level "Advanced". *Collection of scientific and educational works,* 182-198.

Shmatko, N., & Volkova, G. (2023). *Knowledge of foreign languages by Russian scientists*. https://issek.hse.ru/news/226574939.html

Shor, G. A., & Galiullina, A. R. (2021). The role of the English Language in the international cooperation of landscape architects. *World of science, culture, education, 3*(88), 414-416.

Shulman, L. S. (1987). Knowledge and teaching: Foundations of the new reform. *Harvard Educational Review, 57*(1), 1–22. doi:10.17763/haer.57.1.j463w79r56455411

Sinyansky, I. A. (2014). *Typology of real estate: textbook for universities*. Academy.

Slezko, V. V. (2013). *Land Management and land use management*. Infra-M. doi:10.12737/1680

Spada, N., & Lightbown, P. M. (1999). Instruction, first language influence, and developmental readiness in second language acquisition. *Modern Language Journal, 83*(1), 1–22. doi:10.1111/0026-7902.00002

Suhono, S. (2016). Surface strategy taxonomy on the EFL students composition a study of error analysis. *Jurnal Iqra'. Kajian Ilmu Pendidikan, 1*(2), 1–30.

Sun, R. (2017). Error analysis in English writing of Engineering College students based on Pigai.org. *Proceedings of the International Conference on Education Innovation and Social Science (ICEISS 2017)*. 10.2991/iceiss-17.2017.45

Suprunov, S. E. (2019). Studying foreign languages in a non-linguistic university. Actual problems of the general theory of language, translation, intercultural communication and methods of teaching foreign languages, 122-126.

Takanova, O. V. (2008). Vocational-oriented teaching of a foreign language in non-linguistic universities. *Agroengineering*, (6), 49–51.

Taylor, B. P. (1975). Adult language learning strategies and their pedagogical implications. *TESOL Quarterly, 9*(4), 391–399. doi:10.2307/3585623

TemplatesR. (2023). https://uptowork.com/resume-templates#1

Ting, S. H., Mahadhir, M., & Chang, S. L. (2010). Grammatical Errors In Spoken English Of University Students In Oral Communication Course. *GEMA Online™ Journal of Language Studies, 10*, 53–70.

Tokareva, O. S., Alshaibi, A. D. A., & Pasko, O. A. (2021). *Assessment of restoration dynamics of burnt forest area vegetation using landsat satellite data*. Bulletin of the Tomsk Polytechnic University. *Geo Assets Engineering., 332*(7), 191–199.

Touchie, H. Y. (1986). Second language learning errors their types, causes and treatment. *JALT Journal, 8*(1), 75–80.

Transcription, pronunciation and translation of English words online. (2023). https://myefe.ru/anglijskaya-transkriptsiya.html

Trubina, Z. I. (2016). Written translation as a means of teaching a foreign language in a secondary school. *Pedagogical education in Russia*, (3), 171-175.

Tsarevskaya, I. V., & Volodina, M. S. (2016). Dictation as a valuable means of teaching a foreign language. *Modern High Technologies, 8*(1), 153-157.

Use Harvard Referencing Style. (2023). https://curve.coventry.ac.uk/open/file/bdfb947c-9d43-48d3-8ec8-f511682e1dd1/1/The%20CU%20Guide%20to%20Referencing%20in%20Harvard%20Style.pdf

Uvarova, I. V., & Erina, T. F. (2015). Interdisciplinary connections in teaching foreign languages in higher education. *International Journal of Applied and Basic Research*, (3-4), 724–726.

Uzdu Yıldız, F., & Çetin, B. (2020). Errors in written expressions of learners of Turkish as a foreign language: A systematic review. *Journal of Language and Linguistic Studies, 16*(2), 612–625. doi:10.17263/jlls.759261

Van Dijk, T. A. (1989). *Language. Cognition. Communication.* Progress.

Varlamov, A. A., & Galchenko, S. A. (2012). *State real estate cadastre*. Koloss.

Vélez-Rendón, G. (2002). Second language teacher education: A review of the literature. *Foreign Language Annals, 35*(4), 457–467. doi:10.1111/j.1944-9720.2002.tb01884.x

Velliaris, D. M. (2016). Potato, Pot-Ar-To. Tomato, Tom-Ar-To: Is teacher quality and teaching quality the same thing? In D. Polly & C. L. Martin (Eds.), *Handbook of research on teacher education and professional development* (pp. 337–355). IGI Global.

Vitae, C. (CV) Samples, Templates and Writing Tips (2023). https://zety.com/blog/cv-format

Volkova, N. P., Danko, O. Y., & Omelchuk, K. I. (2020). Formation of students' independent activity skills in English lessons through the use of CSR technology. Learning independence of the individual is the basis of lifelong education, 49-54.

Vygotsky, L. S. (1978). *Mind in society: The development of higher psychological processes.* Harvard University Press.

Vygotsky, L. S. (1986). *Thought and language*. MIT Press.

Walger, J. W. (2015). Real estate as an object of capital investment. Problems of market economy management interregional collection of scientific works. Tomsk Polytechnic University (TPU).

Wang, C., & Winstead, L. (2016). *Handbook of Research on Foreign Language Education in the Digital Age*. IGI Global. doi:10.4018/978-1-5225-0177-0

White, L. (1977). Error Analysis and Error Correction in Adult Learners of English as a Second Language. *Working Papers in Bilingualism*, 13.

White, L. (1987). Error Analysis and Error Correction in Adult Learners of English as a Second Language. *Working Papers on Bilingualism*, *13*, 42–58.

Widya Hapsari, E. W. (2021). Error Analysis in Writing Descriptive Text of Pre-Service English Teachers at Kutai Kartanegara University. *Intelegensia: Jurnal Pendidikan Dan Pembelajaran*, *5*(1), 1–12.

Wilson, K., & Devereux, L. (2014). Scaffolding theory: High challenge, high support in Academic Language and Learning (ALL) contexts. *Journal of Academic Language and Learning*, *8*(3), 91–100. https://journal.aall.org.au/index.php/jall/article/download/353/210

Yin, R. K. (2003). *Case study research: Design and methods*. Sage.

Yusuf, Y. Q., Mustafa, F., & Iqbal, R. M. (2021). An inquiry into grammatical errors in writing committed by high achieving EFL students. *International Journal of Language Studies, 15*(2).

Zhakhanova, Z. R. (2015). Structural and Semantic Features of Multicomponent Terms in Modern Russian and English on the Material of the Sublanguage of Architecture and Construction. *Problems of History, Philology, Culture, 2*(48), 274-279.

Zhou, Y., & Wei, M. C. (2018). Strategies in technology-enhanced language learning. *Studies in Second Language Learning and Teaching*, *8*(2), 471–495. doi:10.14746sllt.2018.8.2.13

Related References

To continue our tradition of advancing academic research, we have compiled a list of recommended IGI Global readings. These references will provide additional information and guidance to further enrich your knowledge and assist you with your own research and future publications.

Aburezeq, I. M., & Dweikat, F. F. (2017). Cloud Applications in Language Teaching: Examining Pre-Service Teachers' Expertise, Perceptions and Integration. *International Journal of Distance Education Technologies, 15*(4), 39–60. doi:10.4018/IJDET.2017100103

Acharjya, B., & Das, S. (2022). Adoption of E-Learning During the COVID-19 Pandemic: The Moderating Role of Age and Gender. *International Journal of Web-Based Learning and Teaching Technologies, 17*(2), 1–14. https://doi.org/10.4018/IJWLTT.20220301.oa4

Adams, J. L., & Thomas, S. K. (2022). Non-Linear Curriculum Experiences for Student Learning and Work Design: What Is the Maximum Potential of a Chat Bot? In S. Ramlall, T. Cross, & M. Love (Eds.), *Handbook of Research on Future of Work and Education: Implications for Curriculum Delivery and Work Design* (pp. 299–306). IGI Global. https://doi.org/10.4018/978-1-7998-8275-6.ch018

Adera, B. (2017). Supporting Language and Literacy Development for English Language Learners. In J. Keengwe (Ed.), *Handbook of Research on Promoting Cross-Cultural Competence and Social Justice in Teacher Education* (pp. 339–354). Hershey, PA: IGI Global. doi:10.4018/978-1-5225-0897-7.ch018

Ahamer, G. (2017). Quality Assurance for a Developmental "Global Studies" (GS) Curriculum. In I. Management Association (Ed.), Educational Leadership and Administration: Concepts, Methodologies, Tools, and Applications (pp. 438-477). Hershey, PA: IGI Global. https://doi.org/ doi:10.4018/978-1-5225-1624-8.ch023

Akayoğlu, S., & Seferoğlu, G. (2019). An Analysis of Negotiation of Meaning Functions of Advanced EFL Learners in Second Life: Negotiation of Meaning in Second Life. In M. Kruk (Ed.), *Assessing the Effectiveness of Virtual Technologies in Foreign and Second Language Instruction* (pp. 61–85). IGI Global. https://doi.org/10.4018/978-1-5225-7286-2.ch003

Akella, N. R. (2022). Unravelling the Web of Qualitative Dissertation Writing!: A Student Reflects. In A. Zimmerman (Ed.), *Methodological Innovations in Research and Academic Writing* (pp. 260–282). IGI Global. https://doi.org/10.4018/978-1-7998-8283-1.ch014

Alegre de la Rosa, O. M., & Angulo, L. M. (2017). Social Inclusion and Intercultural Values in a School of Education. In S. Mukerji & P. Tripathi (Eds.), *Handbook of Research on Administration, Policy, and Leadership in Higher Education* (pp. 518–531). Hershey, PA: IGI Global. doi:10.4018/978-1-5225-0672-0.ch020

Alexander, C. (2019). Using Gamification Strategies to Cultivate and Measure Professional Educator Dispositions. *International Journal of Game-Based Learning*, *9*(1), 15–29. https://doi.org/10.4018/IJGBL.2019010102

Anderson, K. M. (2017). Preparing Teachers in the Age of Equity and Inclusion. In I. Management Association (Ed.), Medical Education and Ethics: Concepts, Methodologies, Tools, and Applications (pp. 1532-1554). Hershey, PA: IGI Global. doi:10.4018/978-1-5225-0978-3.ch069

Awdziej, M. (2017). Case Study as a Teaching Method in Marketing. In D. Latusek (Ed.), *Case Studies as a Teaching Tool in Management Education* (pp. 244–263). Hershey, PA: IGI Global. doi:10.4018/978-1-5225-0770-3.ch013

Bakos, J. (2019). Sociolinguistic Factors Influencing English Language Learning. In N. Erdogan & M. Wei (Eds.), *Applied Linguistics for Teachers of Culturally and Linguistically Diverse Learners* (pp. 403–424). IGI Global. https://doi.org/10.4018/978-1-5225-8467-4.ch017

Banas, J. R., & York, C. S. (2017). Pre-Service Teachers' Motivation to Use Technology and the Impact of Authentic Learning Exercises. In L. Tomei (Ed.), *Exploring the New Era of Technology-Infused Education* (pp. 121–140). Hershey, PA: IGI Global. doi:10.4018/978-1-5225-1709-2.ch008

Barton, T. P. (2021). Empowering Educator Allyship by Exploring Racial Trauma and the Disengagement of Black Students. In C. Reneau & M. Villarreal (Eds.), *Handbook of Research on Leading Higher Education Transformation With Social Justice, Equity, and Inclusion* (pp. 186–197). IGI Global. https://doi.org/10.4018/978-1-7998-7152-1.ch013

Benhima, M. (2021). Moroccan English Department Student Attitudes Towards the Use of Distance Education During COVID-19: Moulay Ismail University as a Case Study. *International Journal of Information and Communication Technology Education, 17*(3), 105–122. https://doi.org/10.4018/IJICTE.20210701.oa7

Beycioglu, K., & Wildy, H. (2017). Principal Preparation: The Case of Novice Principals in Turkey. In I. Management Association (Ed.), Educational Leadership and Administration: Concepts, Methodologies, Tools, and Applications (pp. 1152-1169). Hershey, PA: IGI Global. https://doi.org/ doi:10.4018/978-1-5225-1624-8.ch054

Bharwani, S., & Musunuri, D. (2018). Reflection as a Process From Theory to Practice. In M. Khosrow-Pour, D.B.A. (Ed.), Encyclopedia of Information Science and Technology, Fourth Edition (pp. 1529-1539). Hershey, PA: IGI Global. doi:10.4018/978-1-5225-2255-3.ch132

Bhushan, A., Garza, K. B., Perumal, O., Das, S. K., Feola, D. J., Farrell, D., & Birnbaum, A. (2022). Lessons Learned From the COVID-19 Pandemic and the Implications for Pharmaceutical Graduate Education and Research. In C. Ford & K. Garza (Eds.), *Handbook of Research on Updating and Innovating Health Professions Education: Post-Pandemic Perspectives* (pp. 324–345). IGI Global. https://doi.org/10.4018/978-1-7998-7623-6.ch014

Bintz, W., Ciecierski, L. M., & Royan, E. (2021). Using Picture Books With Instructional Strategies to Address New Challenges and Teach Literacy Skills in a Digital World. In L. Haas & J. Tussey (Eds.), *Connecting Disciplinary Literacy and Digital Storytelling in K-12 Education* (pp. 38–58). IGI Global. https://doi.org/10.4018/978-1-7998-5770-9.ch003

Bohjanen, S. L., Cameron-Standerford, A., & Meidl, T. D. (2018). Capacity Building Pedagogy for Diverse Learners. In J. Keengwe (Ed.), *Handbook of Research on Pedagogical Models for Next-Generation Teaching and Learning* (pp. 195–212). Hershey, PA: IGI Global. doi:10.4018/978-1-5225-3873-8.ch011

Brewer, J. C. (2018). Measuring Text Readability Using Reading Level. In M. Khosrow-Pour, D.B.A. (Ed.), Encyclopedia of Information Science and Technology, Fourth Edition (pp. 1499-1507). Hershey, PA: IGI Global. doi:10.4018/978-1-5225-2255-3.ch129

Brookbanks, B. C. (2022). Student Perspectives on Business Education in the USA: Current Attitudes and Necessary Changes in an Age of Disruption. In A. Zhuplev & R. Koepp (Eds.), *Global Trends, Dynamics, and Imperatives for Strategic Development in Business Education in an Age of Disruption* (pp. 214–231). IGI Global. doi:10.4018/978-1-7998-7548-2.ch011

Brown, L. V., Dari, T., & Spencer, N. (2019). Addressing the Impact of Trauma in High Poverty Elementary Schools: An Ecological Model for School Counseling. In K. Daniels & K. Billingsley (Eds.), *Creating Caring and Supportive Educational Environments for Meaningful Learning* (pp. 135–153). IGI Global. https://doi.org/10.4018/978-1-5225-5748-7.ch008

Brown, S. L. (2017). A Case Study of Strategic Leadership and Research in Practice: Principal Preparation Programs that Work – An Educational Administration Perspective of Best Practices for Master's Degree Programs for Principal Preparation. In V. Wang (Ed.), *Encyclopedia of Strategic Leadership and Management* (pp. 1226–1244). Hershey, PA: IGI Global. doi:10.4018/978-1-5225-1049-9.ch086

Brzozowski, M., & Ferster, I. (2017). Educational Management Leadership: High School Principal's Management Style and Parental Involvement in School Management in Israel. In V. Potocan, M. Üngan, & Z. Nedelko (Eds.), *Handbook of Research on Managerial Solutions in Non-Profit Organizations* (pp. 55–74). Hershey, PA: IGI Global. doi:10.4018/978-1-5225-0731-4.ch003

Cahapay, M. B. (2020). Delphi Technique in the Development of Emerging Contents in High School Science Curriculum. *International Journal of Curriculum Development and Learning Measurement, 1*(2), 1–9. https://doi.org/10.4018/IJCDLM.2020070101

Camacho, L. F., & Leon Guerrero, A. E. (2022). Indigenous Student Experience in Higher Education: Implementation of Culturally Sensitive Support. In P. Pangelinan & T. McVey (Eds.), *Learning and Reconciliation Through Indigenous Education in Oceania* (pp. 254–266). IGI Global. https://doi.org/10.4018/978-1-7998-7736-3.ch016

Cannaday, J. (2017). The Masking Effect: Hidden Gifts and Disabilities of 2e Students. In P. Dickenson, P. Keough, & J. Courduff (Eds.), *Preparing Pre-Service Teachers for the Inclusive Classroom* (pp. 220–231). Hershey, PA: IGI Global. doi:10.4018/978-1-5225-1753-5.ch011

Cederquist, S., Fishman, B., & Teasley, S. D. (2022). What's Missing From the College Transcript?: How Employers Make Sense of Student Skills. In Y. Huang (Ed.), *Handbook of Research on Credential Innovations for Inclusive Pathways to Professions* (pp. 234–253). IGI Global. https://doi.org/10.4018/978-1-7998-3820-3.ch012

Cockrell, P., & Gibson, T. (2019). The Untold Stories of Black and Brown Student Experiences in Historically White Fraternities and Sororities. In P. Hoffman-Miller, M. James, & D. Hermond (Eds.), *African American Suburbanization and the Consequential Loss of Identity* (pp. 153–171). IGI Global. https://doi.org/10.4018/978-1-5225-7835-2.ch009

Cohen, M. (2022). Leveraging Content Creation to Boost Student Engagement. In T. Driscoll III, (Ed.), *Designing Effective Distance and Blended Learning Environments in K-12* (pp. 223–239). IGI Global. https://doi.org/10.4018/978-1-7998-6829-3.ch013

Contreras, E. C., & Contreras, I. I. (2018). Development of Communication Skills through Auditory Training Software in Special Education. In M. Khosrow-Pour, D.B.A. (Ed.), Encyclopedia of Information Science and Technology, Fourth Edition (pp. 2431-2441). Hershey, PA: IGI Global. doi:10.4018/978-1-5225-2255-3.ch212

Cooke, L., Schugar, J., Schugar, H., Penny, C., & Bruning, H. (2020). Can Everyone Code?: Preparing Teachers to Teach Computer Languages as a Literacy. In J. Mitchell & E. Vaughn (Eds.), *Participatory Literacy Practices for P-12 Classrooms in the Digital Age* (pp. 163–183). IGI Global. https://doi.org/10.4018/978-1-7998-0000-2.ch009

Cooley, D., & Whitten, E. (2017). Special Education Leadership and the Implementation of Response to Intervention. In F. Topor (Ed.), *Handbook of Research on Individualism and Identity in the Globalized Digital Age* (pp. 265–286). Hershey, PA: IGI Global. doi:10.4018/978-1-5225-0522-8.ch012

Cosner, S., Tozer, S., & Zavitkovsky, P. (2017). Enacting a Cycle of Inquiry Capstone Research Project in Doctoral-Level Leadership Preparation. In I. Management Association (Ed.), Educational Leadership and Administration: Concepts, Methodologies, Tools, and Applications (pp. 1460-1481). Hershey, PA: IGI Global. doi:10.4018/978-1-5225-1624-8.ch067

Crawford, C. M. (2018). Instructional Real World Community Engagement. In M. Khosrow-Pour, D.B.A. (Ed.), Encyclopedia of Information Science and Technology, Fourth Edition (pp. 1474-1486). Hershey, PA: IGI Global. doi:10.4018/978-1-5225-2255-3.ch127

Crosby-Cooper, T., & Pacis, D. (2017). Implementing Effective Student Support Teams. In P. Dickenson, P. Keough, & J. Courduff (Eds.), *Preparing Pre-Service Teachers for the Inclusive Classroom* (pp. 248–262). Hershey, PA: IGI Global. doi:10.4018/978-1-5225-1753-5.ch013

Curran, C. M., & Hawbaker, B. W. (2017). Cultivating Communities of Inclusive Practice: Professional Development for Educators – Research and Practice. In C. Curran & A. Petersen (Eds.), *Handbook of Research on Classroom Diversity and Inclusive Education Practice* (pp. 120–153). Hershey, PA: IGI Global. doi:10.4018/978-1-5225-2520-2.ch006

Dass, S., & Dabbagh, N. (2018). Faculty Adoption of 3D Avatar-Based Virtual World Learning Environments: An Exploratory Case Study. In I. Management Association (Ed.), Technology Adoption and Social Issues: Concepts, Methodologies, Tools, and Applications (pp. 1000-1033). Hershey, PA: IGI Global. https://doi.org/ doi:10.4018/978-1-5225-5201-7.ch045

Davison, A. M., & Scholl, K. G. (2017). Inclusive Recreation as Part of the IEP Process. In C. Curran & A. Petersen (Eds.), *Handbook of Research on Classroom Diversity and Inclusive Education Practice* (pp. 311–330). Hershey, PA: IGI Global. doi:10.4018/978-1-5225-2520-2.ch013

DeCoito, I. (2018). Addressing Digital Competencies, Curriculum Development, and Instructional Design in Science Teacher Education. In M. Khosrow-Pour, D.B.A. (Ed.), Encyclopedia of Information Science and Technology, Fourth Edition (pp. 1420-1431). Hershey, PA: IGI Global. https://doi.org/ doi:10.4018/978-1-5225-2255-3.ch122

DeCoito, I., & Richardson, T. (2017). Beyond Angry Birds™: Using Web-Based Tools to Engage Learners and Promote Inquiry in STEM Learning. In I. Levin & D. Tsybulsky (Eds.), *Digital Tools and Solutions for Inquiry-Based STEM Learning* (pp. 166–196). Hershey, PA: IGI Global. doi:10.4018/978-1-5225-2525-7.ch007

Delmas, P. M. (2017). Research-Based Leadership for Next-Generation Leaders. In R. Styron Jr & J. Styron (Eds.), *Comprehensive Problem-Solving and Skill Development for Next-Generation Leaders* (pp. 1–39). Hershey, PA: IGI Global. doi:10.4018/978-1-5225-1968-3.ch001

Demiray, U., & Ekren, G. (2018). Administrative-Related Evaluation for Distance Education Institutions in Turkey. In K. Buyuk, S. Kocdar, & A. Bozkurt (Eds.), *Administrative Leadership in Open and Distance Learning Programs* (pp. 263–288). Hershey, PA: IGI Global. doi:10.4018/978-1-5225-2645-2.ch011

Dickenson, P. (2017). What do we Know and Where Can We Grow?: Teachers Preparation for the Inclusive Classroom. In P. Dickenson, P. Keough, & J. Courduff (Eds.), *Preparing Pre-Service Teachers for the Inclusive Classroom* (pp. 1–22). Hershey, PA: IGI Global. doi:10.4018/978-1-5225-1753-5.ch001

Ding, Q., & Zhu, H. (2021). Flipping the Classroom in STEM Education. In J. Keengwe (Ed.), *Handbook of Research on Innovations in Non-Traditional Educational Practices* (pp. 155–173). IGI Global. https://doi.org/10.4018/978-1-7998-4360-3.ch008

Dixon, T., & Christison, M. (2021). Teaching English Grammar in a Hybrid Academic ESL Course: A Mixed Methods Study. In K. Kelch, P. Byun, S. Safavi, & S. Cervantes (Eds.), *CALL Theory Applications for Online TESOL Education* (pp. 229–251). IGI Global. https://doi.org/10.4018/978-1-7998-6609-1.ch010

Donne, V., & Hansen, M. (2017). Teachers' Use of Assistive Technologies in Education. In L. Tomei (Ed.), *Exploring the New Era of Technology-Infused Education* (pp. 86–101). Hershey, PA: IGI Global. doi:10.4018/978-1-5225-1709-2.ch006

Donne, V., & Hansen, M. A. (2018). Business and Technology Educators: Practices for Inclusion. In I. Management Association (Ed.), Business Education and Ethics: Concepts, Methodologies, Tools, and Applications (pp. 471-484). Hershey, PA: IGI Global. https://doi.org/ doi:10.4018/978-1-5225-3153-1.ch026

Dos Santos, L. M. (2022). Completing Student-Teaching Internships Online: Instructional Changes During the COVID-19 Pandemic. In M. Alaali (Ed.), *Assessing University Governance and Policies in Relation to the COVID-19 Pandemic* (pp. 106–127). IGI Global. https://doi.org/10.4018/978-1-7998-8279-4.ch007

Dreon, O., Shettel, J., & Bower, K. M. (2017). Preparing Next Generation Elementary Teachers for the Tools of Tomorrow. In M. Grassetti & S. Brookby (Eds.), *Advancing Next-Generation Teacher Education through Digital Tools and Applications* (pp. 143–159). Hershey, PA: IGI Global. doi:10.4018/978-1-5225-0965-3.ch008

Durak, H. Y., & Güyer, T. (2018). Design and Development of an Instructional Program for Teaching Programming Processes to Gifted Students Using Scratch. In J. Cannaday (Ed.), *Curriculum Development for Gifted Education Programs* (pp. 61–99). Hershey, PA: IGI Global. doi:10.4018/978-1-5225-3041-1.ch004

Egorkina, E., Ivanov, M., & Valyavskiy, A. Y. (2018). Students' Research Competence Formation of the Quality of Open and Distance Learning. In V. Mkrttchian & L. Belyanina (Eds.), *Handbook of Research on Students' Research Competence in Modern Educational Contexts* (pp. 364–384). Hershey, PA: IGI Global. doi:10.4018/978-1-5225-3485-3.ch019

Ekren, G., Karataş, S., & Demiray, U. (2017). Understanding of Leadership in Distance Education Management. In I. Management Association (Ed.), Educational Leadership and Administration: Concepts, Methodologies, Tools, and Applications (pp. 34-50). Hershey, PA: IGI Global. https://doi.org/ doi:10.4018/978-1-5225-1624-8.ch003

Elmore, W. M., Young, J. K., Harris, S., & Mason, D. (2017). The Relationship between Individual Student Attributes and Online Course Completion. In K. Shelton & K. Pedersen (Eds.), *Handbook of Research on Building, Growing, and Sustaining Quality E-Learning Programs* (pp. 151–173). Hershey, PA: IGI Global. doi:10.4018/978-1-5225-0877-9.ch008

Ercegovac, I. R., Alfirević, N., & Koludrović, M. (2017). School Principals' Communication and Co-Operation Assessment: The Croatian Experience. In I. Management Association (Ed.), Educational Leadership and Administration: Concepts, Methodologies, Tools, and Applications (pp. 1568-1589). Hershey, PA: IGI Global. https://doi.org/ doi:10.4018/978-1-5225-1624-8.ch072

Everhart, D., & Seymour, D. M. (2017). Challenges and Opportunities in the Currency of Higher Education. In K. Rasmussen, P. Northrup, & R. Colson (Eds.), *Handbook of Research on Competency-Based Education in University Settings* (pp. 41–65). Hershey, PA: IGI Global. doi:10.4018/978-1-5225-0932-5.ch003

Farmer, L. S. (2017). Managing Portable Technologies for Special Education. In V. Wang (Ed.), *Encyclopedia of Strategic Leadership and Management* (pp. 977–987). Hershey, PA: IGI Global. doi:10.4018/978-1-5225-1049-9.ch068

Farmer, L. S. (2018). Optimizing OERs for Optimal ICT Literacy in Higher Education. In J. Keengwe (Ed.), *Handbook of Research on Mobile Technology, Constructivism, and Meaningful Learning* (pp. 366–390). Hershey, PA: IGI Global. doi:10.4018/978-1-5225-3949-0.ch020

Ferguson, B. T. (2019). Supporting Affective Development of Children With Disabilities Through Moral Dilemmas. In S. Ikuta (Ed.), *Handmade Teaching Materials for Students With Disabilities* (pp. 253–275). IGI Global. doi:10.4018/978-1-5225-6240-5.ch011

Fındık, L. Y. (2017). Self-Assessment of Principals Based on Leadership in Complexity. In I. Management Association (Ed.), Educational Leadership and Administration: Concepts, Methodologies, Tools, and Applications (pp. 978-991). Hershey, PA: IGI Global. https://doi.org/ doi:10.4018/978-1-5225-1624-8.ch047

Flor, A. G., & Gonzalez-Flor, B. (2018). Dysfunctional Digital Demeanors: Tales From (and Policy Implications of) eLearning's Dark Side. In I. Management Association (Ed.), The Dark Web: Breakthroughs in Research and Practice (pp. 37-50). Hershey, PA: IGI Global. https://doi.org/ doi:10.4018/978-1-5225-3163-0.ch003

Floyd, K. K., & Shambaugh, N. (2017). Instructional Design for Simulations in Special Education Virtual Learning Spaces. In T. Kidd & L. Morris Jr., (Eds.), *Handbook of Research on Instructional Systems and Educational Technology* (pp. 202–215). Hershey, PA: IGI Global. doi:10.4018/978-1-5225-2399-4.ch018

Freeland, S. F. (2020). Community Schools: Improving Academic Achievement Through Meaningful Engagement. In R. Kronick (Ed.), *Emerging Perspectives on Community Schools and the Engaged University* (pp. 132–144). IGI Global. https://doi.org/10.4018/978-1-7998-0280-8.ch008

Ghanbarzadeh, R., & Ghapanchi, A. H. (2019). Applied Areas of Three Dimensional Virtual Worlds in Learning and Teaching: A Review of Higher Education. In I. Management Association (Ed.), *Virtual Reality in Education: Breakthroughs in Research and Practice* (pp. 172-192). IGI Global. https://doi.org/10.4018/978-1-5225-8179-6.ch008

Giovannini, J. M. (2017). Technology Integration in Preservice Teacher Education Programs: Research-based Recommendations. In M. Grassetti & S. Brookby (Eds.), *Advancing Next-Generation Teacher Education through Digital Tools and Applications* (pp. 82–102). Hershey, PA: IGI Global. doi:10.4018/978-1-5225-0965-3.ch005

Good, S., & Clarke, V. B. (2017). An Integral Analysis of One Urban School System's Efforts to Support Student-Centered Teaching. In J. Keengwe & G. Onchwari (Eds.), *Handbook of Research on Learner-Centered Pedagogy in Teacher Education and Professional Development* (pp. 45–68). Hershey, PA: IGI Global. doi:10.4018/978-1-5225-0892-2.ch003

Guetzoian, E. (2022). Gamification Strategies for Higher Education Student Worker Training. In C. Lane (Ed.), *Handbook of Research on Acquiring 21st Century Literacy Skills Through Game-Based Learning* (pp. 164–179). IGI Global. https://doi.org/10.4018/978-1-7998-7271-9.ch009

Hamidi, F., Owuor, P. M., Hynie, M., Baljko, M., & McGrath, S. (2017). Potentials of Digital Assistive Technology and Special Education in Kenya. In C. Ayo & V. Mbarika (Eds.), *Sustainable ICT Adoption and Integration for Socio-Economic Development* (pp. 125–151). Hershey, PA: IGI Global. doi:10.4018/978-1-5225-2565-3.ch006

Hamim, T., Benabbou, F., & Sael, N. (2022). Student Profile Modeling Using Boosting Algorithms. *International Journal of Web-Based Learning and Teaching Technologies*, *17*(5), 1–13. https://doi.org/10.4018/IJWLTT.20220901.oa4

Henderson, L. K. (2017). Meltdown at Fukushima: Global Catastrophic Events, Visual Literacy, and Art Education. In R. Shin (Ed.), *Convergence of Contemporary Art, Visual Culture, and Global Civic Engagement* (pp. 80–99). Hershey, PA: IGI Global. doi:10.4018/978-1-5225-1665-1.ch005

Hudgins, T., & Holland, J. L. (2018). Digital Badges: Tracking Knowledge Acquisition Within an Innovation Framework. In I. Management Association (Ed.), Wearable Technologies: Concepts, Methodologies, Tools, and Applications (pp. 1118-1132). Hershey, PA: IGI Global. https://doi.org/ doi:10.4018/978-1-5225-5484-4.ch051

Hwang, R., Lin, H., Sun, J. C., & Wu, J. (2019). Improving Learning Achievement in Science Education for Elementary School Students via Blended Learning. *International Journal of Online Pedagogy and Course Design*, *9*(2), 44–62. https://doi.org/10.4018/IJOPCD.2019040104

Jančec, L., & Vodopivec, J. L. (2019). The Implicit Pedagogy and the Hidden Curriculum in Postmodern Education. In J. Vodopivec, L. Jančec, & T. Štemberger (Eds.), *Implicit Pedagogy for Optimized Learning in Contemporary Education* (pp. 41–59). IGI Global. https://doi.org/10.4018/978-1-5225-5799-9.ch003

Janus, M., & Siddiqua, A. (2018). Challenges for Children With Special Health Needs at the Time of Transition to School. In I. Management Association (Ed.), Autism Spectrum Disorders: Breakthroughs in Research and Practice (pp. 339-371). Hershey, PA: IGI Global. doi:10.4018/978-1-5225-3827-1.ch018

Jesus, R. A. (2018). Screencasts and Learning Styles. In M. Khosrow-Pour, D.B.A. (Ed.), Encyclopedia of Information Science and Technology, Fourth Edition (pp. 1548-1558). Hershey, PA: IGI Global. doi:10.4018/978-1-5225-2255-3.ch134

John, G., Francis, N., & Santhakumar, A. B. (2022). Student Engagement: Past, Present, and Future. In S. Ramlall, T. Cross, & M. Love (Eds.), *Handbook of Research on Future of Work and Education: Implications for Curriculum Delivery and Work Design* (pp. 329–341). IGI Global. https://doi.org/10.4018/978-1-7998-8275-6.ch020

Karpinski, A. C., D'Agostino, J. V., Williams, A. K., Highland, S. A., & Mellott, J. A. (2018). The Relationship Between Online Formative Assessment and State Test Scores Using Multilevel Modeling. In M. Khosrow-Pour, D.B.A. (Ed.), Encyclopedia of Information Science and Technology, Fourth Edition (pp. 5183-5192). Hershey, PA: IGI Global. doi:10.4018/978-1-5225-2255-3.ch450

Kats, Y. (2017). Educational Leadership and Integrated Support for Students with Autism Spectrum Disorders. In I. Management Association (Ed.), Educational Leadership and Administration: Concepts, Methodologies, Tools, and Applications (pp. 101-114). Hershey, PA: IGI Global. https://doi.org/ doi:10.4018/978-1-5225-1624-8.ch007

Kaya, G., & Altun, A. (2018). Educational Ontology Development. In M. Khosrow-Pour, D.B.A. (Ed.), Encyclopedia of Information Science and Technology, Fourth Edition (pp. 1441-1450). Hershey, PA: IGI Global. doi:10.4018/978-1-5225-2255-3.ch124

Keough, P. D., & Pacis, D. (2017). Best Practices Implementing Special Education Curriculum and Common Core State Standards using UDL. In P. Dickenson, P. Keough, & J. Courduff (Eds.), *Preparing Pre-Service Teachers for the Inclusive Classroom* (pp. 107–123). Hershey, PA: IGI Global. doi:10.4018/978-1-5225-1753-5.ch006

Kilburn, M., Henckell, M., & Starrett, D. (2018). Factors Contributing to the Effectiveness of Online Students and Instructors. In M. Khosrow-Pour, D.B.A. (Ed.), Encyclopedia of Information Science and Technology, Fourth Edition (pp. 1451-1462). Hershey, PA: IGI Global. doi:10.4018/978-1-5225-2255-3.ch125

Koban Koç, D. (2021). Gender and Language: A Sociolinguistic Analysis of Second Language Writing. In E. Hancı-Azizoglu & N. Kavaklı (Eds.), *Futuristic and Linguistic Perspectives on Teaching Writing to Second Language Students* (pp. 161–177). IGI Global. https://doi.org/10.4018/978-1-7998-6508-7.ch010

Konecny, L. T. (2017). Hybrid, Online, and Flipped Classrooms in Health Science: Enhanced Learning Environments. In I. Management Association (Ed.), Flipped Instruction: Breakthroughs in Research and Practice (pp. 355-370). Hershey, PA: IGI Global. https://doi.org/ doi:10.4018/978-1-5225-1803-7.ch020

Kupietz, K. D. (2021). Gaming and Simulation in Public Education: Teaching Others to Help Themselves and Their Neighbors. In N. Drumhiller, T. Wilkin, & K. Srba (Eds.), *Simulation and Game-Based Learning in Emergency and Disaster Management* (pp. 41–62). IGI Global. https://doi.org/10.4018/978-1-7998-4087-9.ch003

Kwee, C. T. (2022). Assessing the International Student Enrolment Strategies in Australian Universities: A Case Study During the COVID-19 Pandemic. In M. Alaali (Ed.), *Assessing University Governance and Policies in Relation to the COVID-19 Pandemic* (pp. 162–188). IGI Global. https://doi.org/10.4018/978-1-7998-8279-4.ch010

Lauricella, S., & McArthur, F. A. (2022). Taking a Student-Centred Approach to Alternative Digital Credentials: Multiple Pathways Toward the Acquisition of Microcredentials. In D. Piedra (Ed.), *Innovations in the Design and Application of Alternative Digital Credentials* (pp. 57–69). IGI Global. https://doi.org/10.4018/978-1-7998-7697-7.ch003

Llamas, M. F. (2019). Intercultural Awareness in Teaching English for Early Childhood: A Film-Based Approach. In E. Domínguez Romero, J. Bobkina, & S. Stefanova (Eds.), *Teaching Literature and Language Through Multimodal Texts* (pp. 54–68). IGI Global. https://doi.org/10.4018/978-1-5225-5796-8.ch004

Lokhtina, I., & Kkese, E. T. (2022). Reflecting and Adapting to an Academic Workplace Before and After the Lockdown in Greek-Speaking Cyprus: Opportunities and Challenges. In A. Zhuplev & R. Koepp (Eds.), *Global Trends, Dynamics, and Imperatives for Strategic Development in Business Education in an Age of Disruption* (pp. 126–148). IGI Global. https://doi.org/10.4018/978-1-7998-7548-2.ch007

Lovell, K. L. (2017). Development and Evaluation of Neuroscience Computer-Based Modules for Medical Students: Instructional Design Principles and Effectiveness. In J. Stefaniak (Ed.), *Advancing Medical Education Through Strategic Instructional Design* (pp. 262–276). Hershey, PA: IGI Global. doi:10.4018/978-1-5225-2098-6.ch013

Maher, D. (2019). The Use of Course Management Systems in Pre-Service Teacher Education. In J. Keengwe (Ed.), *Handbook of Research on Blended Learning Pedagogies and Professional Development in Higher Education* (pp. 196–213). IGI Global. https://doi.org/10.4018/978-1-5225-5557-5.ch011

Makewa, L. N. (2019). Teacher Technology Competence Base. In L. Makewa, B. Ngussa, & J. Kuboja (Eds.), *Technology-Supported Teaching and Research Methods for Educators* (pp. 247–267). IGI Global. https://doi.org/10.4018/978-1-5225-5915-3.ch014

Mallett, C. A. (2022). School Resource (Police) Officers in Schools: Impact on Campus Safety, Student Discipline, and Learning. In G. Crews (Ed.), *Impact of School Shootings on Classroom Culture, Curriculum, and Learning* (pp. 53–70). IGI Global. https://doi.org/10.4018/978-1-7998-5200-1.ch004

Marinho, J. E., Freitas, I. R., Leão, I. B., Pacheco, L. O., Gonçalves, M. P., Castro, M. J., Silva, P. D., & Moreira, R. J. (2022). Project-Based Learning Application in Higher Education: Student Experiences and Perspectives. In A. Alves & N. van Hattum-Janssen (Eds.), *Training Engineering Students for Modern Technological Advancement* (pp. 146–164). IGI Global. https://doi.org/10.4018/978-1-7998-8816-1.ch007

McCleskey, J. A., & Melton, R. M. (2022). Rolling With the Flow: Online Faculty and Student Presence in a Post-COVID-19 World. In S. Ramlall, T. Cross, & M. Love (Eds.), *Handbook of Research on Future of Work and Education: Implications for Curriculum Delivery and Work Design* (pp. 307–328). IGI Global. https://doi.org/10.4018/978-1-7998-8275-6.ch019

McCormack, V. F., Stauffer, M., Fishley, K., Hohenbrink, J., Mascazine, J. R., & Zigler, T. (2018). Designing a Dual Licensure Path for Middle Childhood and Special Education Teacher Candidates. In D. Polly, M. Putman, T. Petty, & A. Good (Eds.), *Innovative Practices in Teacher Preparation and Graduate-Level Teacher Education Programs* (pp. 21–36). Hershey, PA: IGI Global. doi:10.4018/978-1-5225-3068-8.ch002

McDaniel, R. (2017). Strategic Leadership in Instructional Design: Applying the Principles of Instructional Design through the Lens of Strategic Leadership to Distance Education. In V. Wang (Ed.), *Encyclopedia of Strategic Leadership and Management* (pp. 1570–1584). Hershey, PA: IGI Global. doi:10.4018/978-1-5225-1049-9.ch109

McKinney, R. E., Halli-Tierney, A. D., Gold, A. E., Allen, R. S., & Carroll, D. G. (2022). Interprofessional Education: Using Standardized Cases in Face-to-Face and Remote Learning Settings. In C. Ford & K. Garza (Eds.), *Handbook of Research on Updating and Innovating Health Professions Education: Post-Pandemic Perspectives* (pp. 24–42). IGI Global. https://doi.org/10.4018/978-1-7998-7623-6.ch002

Meintjes, H. H. (2021). Learner Views of a Facebook Page as a Supportive Digital Pedagogical Tool at a Public South African School in a Grade 12 Business Studies Class. *International Journal of Smart Education and Urban Society, 12*(2), 32–45. https://doi.org/10.4018/IJSEUS.2021040104

Melero-García, F. (2022). Training Bilingual Interpreters in Healthcare Settings: Student Perceptions of Online Learning. In J. LeLoup & P. Swanson (Eds.), *Handbook of Research on Effective Online Language Teaching in a Disruptive Environment* (pp. 288–310). IGI Global. https://doi.org/10.4018/978-1-7998-7720-2.ch015

Meletiadou, E. (2022). The Use of Peer Assessment as an Inclusive Learning Strategy in Higher Education Institutions: Enhancing Student Writing Skills and Motivation. In E. Meletiadou (Ed.), *Handbook of Research on Policies and Practices for Assessing Inclusive Teaching and Learning* (pp. 1–26). IGI Global. https://doi.org/10.4018/978-1-7998-8579-5.ch001

Memon, R. N., Ahmad, R., & Salim, S. S. (2018). Critical Issues in Requirements Engineering Education. In I. Management Association (Ed.), Computer Systems and Software Engineering: Concepts, Methodologies, Tools, and Applications (pp. 1953-1976). Hershey, PA: IGI Global. doi:10.4018/978-1-5225-3923-0.ch081

Mendenhall, R. (2017). Western Governors University: CBE Innovator and National Model. In K. Rasmussen, P. Northrup, & R. Colson (Eds.), *Handbook of Research on Competency-Based Education in University Settings* (pp. 379–400). Hershey, PA: IGI Global. doi:10.4018/978-1-5225-0932-5.ch019

Mense, E. G., Griggs, D. M., & Shanks, J. N. (2018). School Leaders in a Time of Accountability and Data Use: Preparing Our Future School Leaders in Leadership Preparation Programs. In E. Mense & M. Crain-Dorough (Eds.), *Data Leadership for K-12 Schools in a Time of Accountability* (pp. 235–259). Hershey, PA: IGI Global. doi:10.4018/978-1-5225-3188-3.ch012

Mense, E. G., Griggs, D. M., & Shanks, J. N. (2018). School Leaders in a Time of Accountability and Data Use: Preparing Our Future School Leaders in Leadership Preparation Programs. In E. Mense & M. Crain-Dorough (Eds.), *Data Leadership for K-12 Schools in a Time of Accountability* (pp. 235–259). Hershey, PA: IGI Global. doi:10.4018/978-1-5225-3188-3.ch012

Mestry, R., & Naicker, S. R. (2017). Exploring Distributive Leadership in South African Public Primary Schools in the Soweto Region. In I. Management Association (Ed.), Educational Leadership and Administration: Concepts, Methodologies, Tools, and Applications (pp. 1041-1064). Hershey, PA: IGI Global. doi:10.4018/978-1-5225-1624-8.ch050

Monaghan, C. H., & Boboc, M. (2017). (Re) Defining Leadership in Higher Education in the U.S. In V. Wang (Ed.), *Encyclopedia of Strategic Leadership and Management* (pp. 567–579). Hershey, PA: IGI Global. doi:10.4018/978-1-5225-1049-9.ch040

Morall, M. B. (2021). Reimagining Mobile Phones: Multiple Literacies and Digital Media Compositions. In C. Moran (Eds.), *Affordances and Constraints of Mobile Phone Use in English Language Arts Classrooms* (pp. 41-53). IGI Global. https://doi.org/10.4018/978-1-7998-5805-8.ch003

Mthethwa, V. (2022). Student Governance and the Academic Minefield During COVID-19 Lockdown in South Africa. In M. Alaali (Ed.), *Assessing University Governance and Policies in Relation to the COVID-19 Pandemic* (pp. 255–276). IGI Global. https://doi.org/10.4018/978-1-7998-8279-4.ch015

Muthee, J. M., & Murungi, C. G. (2018). Relationship Among Intelligence, Achievement Motivation, Type of School, and Academic Performance of Kenyan Urban Primary School Pupils. In M. Khosrow-Pour, D.B.A. (Ed.), Encyclopedia of Information Science and Technology, Fourth Edition (pp. 1540-1547). Hershey, PA: IGI Global. https://doi.org/ doi:10.4018/978-1-5225-2255-3.ch133

Naranjo, J. (2018). Meeting the Need for Inclusive Educators Online: Teacher Education in Inclusive Special Education and Dual-Certification. In D. Polly, M. Putman, T. Petty, & A. Good (Eds.), *Innovative Practices in Teacher Preparation and Graduate-Level Teacher Education Programs* (pp. 106–122). Hershey, PA: IGI Global. doi:10.4018/978-1-5225-3068-8.ch007

Nkabinde, Z. P. (2017). Multiculturalism in Special Education: Perspectives of Minority Children in Urban Schools. In J. Keengwe (Ed.), *Handbook of Research on Promoting Cross-Cultural Competence and Social Justice in Teacher Education* (pp. 382–397). Hershey, PA: IGI Global. doi:10.4018/978-1-5225-0897-7.ch020

Nkabinde, Z. P. (2018). Online Instruction: Is the Quality the Same as Face-to-Face Instruction? In J. Keengwe (Ed.), *Handbook of Research on Digital Content, Mobile Learning, and Technology Integration Models in Teacher Education* (pp. 300–314). Hershey, PA: IGI Global. doi:10.4018/978-1-5225-2953-8.ch016

Nugroho, A., & Albusaidi, S. S. (2022). Internationalization of Higher Education: The Methodological Critiques on the Research Related to Study Overseas and International Experience. In H. Magd & S. Kunjumuhammed (Eds.), *Global Perspectives on Quality Assurance and Accreditation in Higher Education Institutions* (pp. 75–89). IGI Global. https://doi.org/10.4018/978-1-7998-8085-1.ch005

Nulty, Z., & West, S. G. (2022). Student Engagement and Supporting Students With Accommodations. In P. Bull & G. Patterson (Eds.), *Redefining Teacher Education and Teacher Preparation Programs in the Post-COVID-19 Era* (pp. 99–116). IGI Global. https://doi.org/10.4018/978-1-7998-8298-5.ch006

O'Connor, J. R. Jr, & Jackson, K. N. (2017). The Use of iPad® Devices and "Apps" for ASD Students in Special Education and Speech Therapy. In Y. Kats (Ed.), *Supporting the Education of Children with Autism Spectrum Disorders* (pp. 267–283). Hershey, PA: IGI Global. doi:10.4018/978-1-5225-0816-8.ch014

Okolie, U. C., & Yasin, A. M. (2017). TVET in Developing Nations and Human Development. In U. Okolie & A. Yasin (Eds.), *Technical Education and Vocational Training in Developing Nations* (pp. 1–25). Hershey, PA: IGI Global. doi:10.4018/978-1-5225-1811-2.ch001

Pack, A., & Barrett, A. (2021). A Review of Virtual Reality and English for Academic Purposes: Understanding Where to Start. *International Journal of Computer-Assisted Language Learning and Teaching*, *11*(1), 72–80. https://doi.org/10.4018/IJCALLT.2021010105

Pashollari, E. (2019). Building Sustainability Through Environmental Education: Education for Sustainable Development. In L. Wilson, & C. Stevenson (Eds.), *Building Sustainability Through Environmental Education* (pp. 72-88). IGI Global. https://doi.org/10.4018/978-1-5225-7727-0.ch004

Paulson, E. N. (2017). Adapting and Advocating for an Online EdD Program in Changing Times and "Sacred" Cultures. In I. Management Association (Ed.), Educational Leadership and Administration: Concepts, Methodologies, Tools, and Applications (pp. 1849-1876). Hershey, PA: IGI Global. https://doi.org/doi:10.4018/978-1-5225-1624-8.ch085

Petersen, A. J., Elser, C. F., Al Nassir, M. N., Stakey, J., & Everson, K. (2017). The Year of Teaching Inclusively: Building an Elementary Classroom for All Students. In C. Curran & A. Petersen (Eds.), *Handbook of Research on Classroom Diversity and Inclusive Education Practice* (pp. 332–348). Hershey, PA: IGI Global. doi:10.4018/978-1-5225-2520-2.ch014

Pfannenstiel, K. H., & Sanders, J. (2017). Characteristics and Instructional Strategies for Students With Mathematical Difficulties: In the Inclusive Classroom. In C. Curran & A. Petersen (Eds.), *Handbook of Research on Classroom Diversity and Inclusive Education Practice* (pp. 250–281). Hershey, PA: IGI Global. doi:10.4018/978-1-5225-2520-2.ch011

Phan, A. N. (2022). Quality Assurance of Higher Education From the Glonacal Agency Heuristic: An Example From Vietnam. In H. Magd & S. Kunjumuhammed (Eds.), *Global Perspectives on Quality Assurance and Accreditation in Higher Education Institutions* (pp. 136–155). IGI Global. https://doi.org/10.4018/978-1-7998-8085-1.ch008

Preast, J. L., Bowman, N., & Rose, C. A. (2017). Creating Inclusive Classroom Communities Through Social and Emotional Learning to Reduce Social Marginalization Among Students. In C. Curran & A. Petersen (Eds.), *Handbook of Research on Classroom Diversity and Inclusive Education Practice* (pp. 183–200). Hershey, PA: IGI Global. doi:10.4018/978-1-5225-2520-2.ch008

Randolph, K. M., & Brady, M. P. (2018). Evolution of Covert Coaching as an Evidence-Based Practice in Professional Development and Preparation of Teachers. In V. Bryan, A. Musgrove, & J. Powers (Eds.), *Handbook of Research on Human Development in the Digital Age* (pp. 281–299). Hershey, PA: IGI Global. doi:10.4018/978-1-5225-2838-8.ch013

Rell, A. B., Puig, R. A., Roll, F., Valles, V., Espinoza, M., & Duque, A. L. (2017). Addressing Cultural Diversity and Global Competence: The Dual Language Framework. In L. Leavitt, S. Wisdom, & K. Leavitt (Eds.), *Cultural Awareness and Competency Development in Higher Education* (pp. 111–131). Hershey, PA: IGI Global. doi:10.4018/978-1-5225-2145-7.ch007

Richards, M., & Guzman, I. R. (2020). Academic Assessment of Critical Thinking in Distance Education Information Technology Programs. In I. Management Association (Ed.), *Learning and Performance Assessment: Concepts, Methodologies, Tools, and Applications* (pp. 1-19). IGI Global. https://doi.org/10.4018/978-1-7998-0420-8.ch001

Riel, J., Lawless, K. A., & Brown, S. W. (2017). Defining and Designing Responsive Online Professional Development (ROPD): A Framework to Support Curriculum Implementation. In T. Kidd & L. Morris Jr., (Eds.), *Handbook of Research on Instructional Systems and Educational Technology* (pp. 104–115). Hershey, PA: IGI Global. doi:10.4018/978-1-5225-2399-4.ch010

Roberts, C. (2017). Advancing Women Leaders in Academe: Creating a Culture of Inclusion. In S. Mukerji & P. Tripathi (Eds.), *Handbook of Research on Administration, Policy, and Leadership in Higher Education* (pp. 256–273). Hershey, PA: IGI Global. doi:10.4018/978-1-5225-0672-0.ch012

Rodgers, W. J., Kennedy, M. J., Alves, K. D., & Romig, J. E. (2017). A Multimedia Tool for Teacher Education and Professional Development. In C. Martin & D. Polly (Eds.), *Handbook of Research on Teacher Education and Professional Development* (pp. 285–296). Hershey, PA: IGI Global. doi:10.4018/978-1-5225-1067-3.ch015

Romanowski, M. H. (2017). Qatar's Educational Reform: Critical Issues Facing Principals. In I. Management Association (Ed.), Educational Leadership and Administration: Concepts, Methodologies, Tools, and Applications (pp. 1758-1773). Hershey, PA: IGI Global. https://doi.org/ doi:10.4018/978-1-5225-1624-8.ch080

Ruffin, T. R., Hawkins, D. P., & Lee, D. I. (2018). Increasing Student Engagement and Participation Through Course Methodology. In M. Khosrow-Pour, D.B.A. (Ed.), Encyclopedia of Information Science and Technology, Fourth Edition (pp. 1463-1473). Hershey, PA: IGI Global. doi:10.4018/978-1-5225-2255-3.ch126

Sabina, L. L., Curry, K. A., Harris, E. L., Krumm, B. L., & Vencill, V. (2017). Assessing the Performance of a Cohort-Based Model Using Domestic and International Practices. In I. Management Association (Ed.), Educational Leadership and Administration: Concepts, Methodologies, Tools, and Applications(pp. 913-929). Hershey, PA: IGI Global. https://doi.org/ doi:10.4018/978-1-5225-1624-8.ch044

Samkian, A., Pascarella, J., & Slayton, J. (2022). Towards an Anti-Racist, Culturally Responsive, and LGBTQ+ Inclusive Education: Developing Critically-Conscious Educational Leaders. In E. Cain-Sanschagrin, R. Filback, & J. Crawford (Eds.), *Cases on Academic Program Redesign for Greater Racial and Social Justice* (pp. 150–175). IGI Global. https://doi.org/10.4018/978-1-7998-8463-7.ch007

Santamaría, A. P., Webber, M., & Santamaría, L. J. (2017). Effective School Leadership for Māori Achievement: Building Capacity through Indigenous, National, and International Cross-Cultural Collaboration. In I. Management Association (Ed.), Educational Leadership and Administration: Concepts, Methodologies, Tools, and Applications (pp. 1547-1567). Hershey, PA: IGI Global. https://doi.org/ doi:10.4018/978-1-5225-1624-8.ch071

Santamaría, L. J. (2017). Culturally Responsive Educational Leadership in Cross-Cultural International Contexts. In I. Management Association (Ed.), Educational Leadership and Administration: Concepts, Methodologies, Tools, and Applications (pp. 1380-1400). Hershey, PA: IGI Global. https://doi.org/ doi:10.4018/978-1-5225-1624-8.ch064

Segredo, M. R., Cistone, P. J., & Reio, T. G. (2017). Relationships Between Emotional Intelligence, Leadership Style, and School Culture. *International Journal of Adult Vocational Education and Technology*, *8*(3), 25–43. doi:10.4018/IJAVET.2017070103

Shalev, N. (2017). Empathy and Leadership From the Organizational Perspective. In Z. Nedelko & M. Brzozowski (Eds.), *Exploring the Influence of Personal Values and Cultures in the Workplace* (pp. 348–363). Hershey, PA: IGI Global. doi:10.4018/978-1-5225-2480-9.ch018

Siamak, M., Fathi, S., & Isfandyari-Moghaddam, A. (2018). Assessment and Measurement of Education Programs of Information Literacy. In R. Bhardwaj (Ed.), *Digitizing the Modern Library and the Transition From Print to Electronic* (pp. 164–192). Hershey, PA: IGI Global. doi:10.4018/978-1-5225-2119-8.ch007

Siu, K. W., & García, G. J. (2017). Disruptive Technologies and Education: Is There Any Disruption After All? In I. Management Association (Ed.), Educational Leadership and Administration: Concepts, Methodologies, Tools, and Applications (pp. 757-778). Hershey, PA: IGI Global. https://doi.org/ doi:10.4018/978-1-5225-1624-8.ch037

Slagter van Tryon, P. J. (2017). The Nurse Educator's Role in Designing Instruction and Instructional Strategies for Academic and Clinical Settings. In J. Stefaniak (Ed.), *Advancing Medical Education Through Strategic Instructional Design* (pp. 133–149). Hershey, PA: IGI Global. doi:10.4018/978-1-5225-2098-6.ch006

Slattery, C. A. (2018). Literacy Intervention and the Differentiated Plan of Instruction. In *Developing Effective Literacy Intervention Strategies: Emerging Research and Opportunities* (pp. 41–62). Hershey, PA: IGI Global. doi:10.4018/978-1-5225-5007-5.ch003

Smith, A. R. (2017). Ensuring Quality: The Faculty Role in Online Higher Education. In K. Shelton & K. Pedersen (Eds.), *Handbook of Research on Building, Growing, and Sustaining Quality E-Learning Programs* (pp. 210–231). Hershey, PA: IGI Global. doi:10.4018/978-1-5225-0877-9.ch011

Souders, T. M. (2017). Understanding Your Learner: Conducting a Learner Analysis. In J. Stefaniak (Ed.), *Advancing Medical Education Through Strategic Instructional Design* (pp. 1–29). Hershey, PA: IGI Global. doi:10.4018/978-1-5225-2098-6.ch001

Spring, K. J., Graham, C. R., & Ikahihifo, T. B. (2018). Learner Engagement in Blended Learning. In M. Khosrow-Pour, D.B.A. (Ed.), Encyclopedia of Information Science and Technology, Fourth Edition (pp. 1487-1498). Hershey, PA: IGI Global. doi:10.4018/978-1-5225-2255-3.ch128

Storey, V. A., Anthony, A. K., & Wahid, P. (2017). Gender-Based Leadership Barriers: Advancement of Female Faculty to Leadership Positions in Higher Education. In V. Wang (Ed.), *Encyclopedia of Strategic Leadership and Management* (pp. 244–258). Hershey, PA: IGI Global. doi:10.4018/978-1-5225-1049-9.ch018

Stottlemyer, D. (2018). Develop a Teaching Model Plan for a Differentiated Learning Approach. In *Differentiated Instructional Design for Multicultural Environments: Emerging Research and Opportunities* (pp. 106–130). Hershey, PA: IGI Global. doi:10.4018/978-1-5225-5106-5.ch005

Stottlemyer, D. (2018). Developing a Multicultural Environment. In *Differentiated Instructional Design for Multicultural Environments: Emerging Research and Opportunities* (pp. 1–27). Hershey, PA: IGI Global. doi:10.4018/978-1-5225-5106-5.ch001

Swagerty, T. (2022). Digital Access to Culturally Relevant Curricula: The Impact on the Native and Indigenous Student. In E. Reeves & C. McIntyre (Eds.), *Multidisciplinary Perspectives on Diversity and Equity in a Virtual World* (pp. 99–113). IGI Global. https://doi.org/10.4018/978-1-7998-8028-8.ch006

Swami, B. N., Gobona, T., & Tsimako, J. J. (2017). Academic Leadership: A Case Study of the University of Botswana. In N. Baporikar (Ed.), *Innovation and Shifting Perspectives in Management Education* (pp. 1–32). Hershey, PA: IGI Global. doi:10.4018/978-1-5225-1019-2.ch001

Swanson, K. W., & Collins, G. (2018). Designing Engaging Instruction for the Adult Learners. In M. Khosrow-Pour, D.B.A. (Ed.), Encyclopedia of Information Science and Technology, Fourth Edition (pp. 1432-1440). Hershey, PA: IGI Global. doi:10.4018/978-1-5225-2255-3.ch123

Swartz, B. A., Lynch, J. M., & Lynch, S. D. (2018). Embedding Elementary Teacher Education Coursework in Local Classrooms: Examples in Mathematics and Special Education. In D. Polly, M. Putman, T. Petty, & A. Good (Eds.), *Innovative Practices in Teacher Preparation and Graduate-Level Teacher Education Programs* (pp. 262–292). Hershey, PA: IGI Global. doi:10.4018/978-1-5225-3068-8.ch015

Taliadorou, N., & Pashiardis, P. (2017). Emotional Intelligence and Political Skill Really Matter in Educational Leadership. In I. Management Association (Ed.), Educational Leadership and Administration: Concepts, Methodologies, Tools, and Applications (pp. 1274-1303). Hershey, PA: IGI Global. https://doi.org/doi:10.4018/978-1-5225-1624-8.ch060

Tandoh, K. A., & Ebe-Arthur, J. E. (2018). Effective Educational Leadership in the Digital Age: An Examination of Professional Qualities and Best Practices. In J. Keengwe (Ed.), *Handbook of Research on Digital Content, Mobile Learning, and Technology Integration Models in Teacher Education* (pp. 244–265). Hershey, PA: IGI Global. doi:10.4018/978-1-5225-2953-8.ch013

Tobin, M. T. (2018). Multimodal Literacy. In M. Khosrow-Pour, D.B.A. (Ed.), Encyclopedia of Information Science and Technology, Fourth Edition (pp. 1508-1516). Hershey, PA: IGI Global. doi:10.4018/978-1-5225-2255-3.ch130

Torres, K. M., Arrastia-Chisholm, M. C., & Tackett, S. (2019). A Phenomenological Study of Pre-Service Teachers' Perceptions of Completing ESOL Field Placements. *International Journal of Teacher Education and Professional Development*, *2*(2), 85–101. https://doi.org/10.4018/IJTEPD.2019070106

Torres, M. C., Salamanca, Y. N., Cely, J. P., & Aguilar, J. L. (2020). All We Need is a Boost! Using Multimodal Tools and the Translanguaging Strategy: Strengthening Speaking in the EFL Classroom. *International Journal of Computer-Assisted Language Learning and Teaching*, *10*(3), 28–47. doi:10.4018/IJCALLT.2020070103

Torres, M. L., & Ramos, V. J. (2018). Music Therapy: A Pedagogical Alternative for ASD and ID Students in Regular Classrooms. In P. Epler (Ed.), *Instructional Strategies in General Education and Putting the Individuals With Disabilities Act (IDEA) Into Practice* (pp. 222–244). Hershey, PA: IGI Global. doi:10.4018/978-1-5225-3111-1.ch008

Toulassi, B. (2017). Educational Administration and Leadership in Francophone Africa: 5 Dynamics to Change Education. In S. Mukerji & P. Tripathi (Eds.), *Handbook of Research on Administration, Policy, and Leadership in Higher Education* (pp. 20–45). Hershey, PA: IGI Global. doi:10.4018/978-1-5225-0672-0.ch002

Umair, S., & Sharif, M. M. (2018). Predicting Students Grades Using Artificial Neural Networks and Support Vector Machine. In M. Khosrow-Pour, D.B.A. (Ed.), Encyclopedia of Information Science and Technology, Fourth Edition (pp. 5169-5182). Hershey, PA: IGI Global. doi:10.4018/978-1-5225-2255-3.ch449

Vettraino, L., Castello, V., Guspini, M., & Guglielman, E. (2018). Self-Awareness and Motivation Contrasting ESL and NEET Using the SAVE System. In M. Khosrow-Pour, D.B.A. (Ed.), Encyclopedia of Information Science and Technology, Fourth Edition (pp. 1559-1568). Hershey, PA: IGI Global. doi:10.4018/978-1-5225-2255-3.ch135

Wiemelt, J. (2017). Critical Bilingual Leadership for Emergent Bilingual Students. In I. Management Association (Ed.), Educational Leadership and Administration: Concepts, Methodologies, Tools, and Applications (pp. 1606-1631). Hershey, PA: IGI Global. doi:10.4018/978-1-5225-1624-8.ch074

Wolf, F., Seyfarth, F. C., & Pflaum, E. (2018). Scalable Capacity-Building for Geographically Dispersed Learners: Designing the MOOC "Sustainable Energy in Small Island Developing States (SIDS)". In U. Pandey & V. Indrakanti (Eds.), *Open and Distance Learning Initiatives for Sustainable Development* (pp. 58–83). Hershey, PA: IGI Global. doi:10.4018/978-1-5225-2621-6.ch003

Woodley, X. M., Mucundanyi, G., & Lockard, M. (2017). Designing Counter-Narratives: Constructing Culturally Responsive Curriculum Online. *International Journal of Online Pedagogy and Course Design*, *7*(1), 43–56. doi:10.4018/IJOPCD.2017010104

Yell, M. L., & Christle, C. A. (2017). The Foundation of Inclusion in Federal Legislation and Litigation. In C. Curran & A. Petersen (Eds.), *Handbook of Research on Classroom Diversity and Inclusive Education Practice* (pp. 27–52). Hershey, PA: IGI Global. doi:10.4018/978-1-5225-2520-2.ch002

Zinner, L. (2019). Fostering Academic Citizenship With a Shared Leadership Approach. In C. Zhu & M. Zayim-Kurtay (Eds.), *University Governance and Academic Leadership in the EU and China* (pp. 99–117). IGI Global. https://doi.org/10.4018/978-1-5225-7441-5.ch007

About the Contributors

Nadezhda Anatolievna Lebedeva is a Doctor of Philosophy in the field of Cultural Studies and works at the Kherson State Agrarian University (Kherson, Ukraine) as the department head of Humanitarian Disciplines Chair, and a professor of the International Personnel Academy in Kyiv and a full member of the Eurasian Academy of Television and Radio. She is the main editor of the international journal "Universum: Philology & Arts", a reviewer and a member of the editorial board of scientific collections of international conferences and scientific journals ("Journal of Social, Humanity, and Education", "Journal of Sustainable Tourism and Entrepreneurship" Goodwood Publishing), the author of more than a hundred publications, including Scopus and WoS, has recommendations and certificates of honour for successful scientific activity). Experienced in project cinematography management and developed lecture courses in a foreign language for students of state administration. She is the author of more than 100 scientific works. The focus of her research is the cultural and social aspects of regional development. N. Lebedeva graduated from the Kherson State Pedagogical Institute in 1996. After graduating from the post-graduate course at the Kherson State University, became a lecturer at the History of World Culture and Literature Department of the Kherson State University. In June 2008 was enrolled in the doctoral program of the Interregional Academy of Personnel Management in Kyiv, specializing in Cultural Studies. She was awarded the medal "For Civil Courage" of the Afghan Veterans' Ukrainian Union for the book "My Afghan" by Nikolay Vishnyak". Participation in the creation of documentary films and scientific articles is marked by membership in the Eurasian Academy of Television and Radio (November 2012). The monograph "Cultural Aspect of Alexander Elagin's Creativeness" was awarded 1st place in scientific direction "Cultural Studies" in 2017 and became the winner of the competition for educational projects "Interclover – 2017" in the nomination – "Research Project". She also was invited to be a member of the Ukrainian Assembly of Doctors of Sciences in the Field of State Administration (February 2018). Scope of her scientific interests: state administration, philosophy, pedagogy, Russian, Ukrainian and English philology, cultural studies and ethnology.

* * *

Isabel María Garcia Conesa has a degree in English Studies from the University of Alicante and a PhD from the National University of Distance Education (UNED). She is currently working as a full time lecturer at the Centro Universitario de la Defensa in San Javier (Spain). She has been awarded a scholarship by the Franklin Institute (University of Alcala de Henares, Spain) and the Radcliffe Institute for Advanced Study (Harvard University, USA), where she conducted a pre-doctoral research stay in the year 2012. Among her main lines of research, we can highlight the role of different women in literature and culture of the United States in contrast to Francophone writers with publication in Spanish journals like Revista Estudios Humanísticos (University of León), Tonos Digital (University of Murcia), Dossiers Feministes (University Jaime I), Prisma Social, Raudem (University of Almeria), Camino Real (University of Alcala de Henres), or Nomadas (University Complutense of Madrid). She also focuses on the study of the history of the teaching of English and gender studies with publications in journals such as Revista Feminismo (University of Alicante) or even Quaderns Digitals.

Antonio D. Juan Rubio has a degree in English Studies from the University of Murcia and a PhD from the National University of Distance Education (UNED) with the positive accreditation by the ANECA body, being given the Extraordinary Doctorate Award. He is currently working as a professor at the University of Granada. He has been awarded a scholarship by University College (Cork, Ireland), the Franklin Institute (University of Alcala de Henares, Spain) and the Radcliffe Institute for Advanced Study (Harvard University, USA), where he conducted a pre-doctoral research visit in the year 2012. He is currently a member of the scientific committee of several national and international journals as well as a member of the editorial board of several international journals. He also belongs to the organizing and scientific committee of several conferences organized by the Athens Institute for Education and Research (Greece). Among his main lines of research we can emphasize the following aspects: cultural studies in the United States; gender issues associated with the role of women in the Anglo-American literature; or the teaching practice and process of English.

Ali Kurt obtained his Ph.D. in English Language Education, focusing on intrinsic motivation for language acquisition, from Canakkale 18Mart University, Turkey, in 2021. Ali Kurt's dissertation significantly contributed to understanding the influential role of intrinsic motivation and has offered innovative solutions to this common issue. With a career spanning over two decades, Dr. Kurt has demonstrated exemplary teaching abilities across various educational settings. He has taught diverse subjects, including special education classes, General English, and Academic English, at primary, secondary, and pre-sessional English schools, as well as many different courses, such as Academic Writing, Communication Skills, and Communicative Competence in different departments. As an 'English exams' expert since 2015, Kurt has successfully prepared students for both national and international exams. Currently, Dr. Kurt holds the position of director at a renowned university's pre-sessional English school, where he oversees operations and leads a team of educators. Ali Kurt's contributions to the academic community extend beyond teaching. He has conducted influential research on learning and motivation, digitized language learning, digital burnout, productive skills, error analysis, and English for special purposes. As a thought leader in the field, he endeavors to advance language education practices.

Zinaida Lebiedieva, Master of Arts, studied at the piano faculty of the Ukrainian National Tchaikovsky Academy of Music (Kyiv, Ukraine) since 2015 (2019 – bachelor's degree with honours, class of acting associate professor S. I. Ryabov). Until 2021, she studied in the class of the acting professor, a candidate in art history of the famous modern Ukrainian composer Oleg Bezborodyko. In 2021 graduated from the Ukrainian National Tchaikovsky Academy of Music and received a master's degree specializing in "Piano". In 2011, she started working as a concertmaster at the State Children's Choreographic Studio of Ukrainian Dance "Barvinochok". From 2013 to 2015, she worked as a concertmaster in various state art institutions of elementary music education in Kyiv. Since 2015, she has been accepted as a concertmaster at the Serzh Lyfar Municipal Academy of Dance. From 2018 to 2019, she combined her main concertmaster activities with the opera studio and from 2020 to 2021 at the Kyiv Municipal Academy of Music named after R. M. Glyer and the Ukrainian National Tchaikovsky Academy of Music. She is also active in concert music as a vocal accompanist of the Ukrainian National Tchaikovsky Academy of Music and conducts active musical concert activities as a vocal accompanist. She has repeatedly performed as a soloist with an orchestra. In 2005, Zinayida was awarded a certificate from the governor of the Kherson region "For active participation in reviews of competitions and significant successes in performance skills" and became a graduate and laureate of regional, all – Ukrainian and international competitions.

Olga Pasko worked at the National Research Tomsk Polytechnic University, Engineering School of Natural Resources, Department of Geology, Professor. Currently, he is Vice-Rector for Research at the National Open Institute St. Petersburg. She has a work experience in the "Agrophysical Research Institute", St. Petersburg as the head of the experimental research sector. Main responsibilities: management of profiles in the areas of "Land Management" (master's degree) and "Land management, cadastre and land monitoring" (postgraduate studies); preparation and implementation of accreditation for these specialities in 2016 and 2021; planning, organization and control of educational, educational and methodological work in the disciplines provided, taking into account the specifics of the contingent of students. Olga Anatolievna created some working curricula for the provided disciplines and management of their development by other teachers. She also conducted all types of training sessions, managing course and diploma projects and research work for bachelor's and masters. She is also known for her teaching experience at universities in Tomsk: National Research Tomsk Polytechnic University, National Research Tomsk State University, Tomsk University of Architecture and Civil Engineering, Tomsk Agricultural Institute – a branch of the Novosibirsk Agrarian University. More than 200 publications have been published, including in journals included in the Scopus and WoS databases – 20. Hirsch Scopus Index – 4, WoS – 3, Hirsch RSCI – 10. Awarded with thanks from the Ministry of Science and Higher Education of the Russian Federation, the Administration of the Tomsk Region, State Duma of the Tomsk Region, Administration of the Novosibirsk Region, Government of the Novosibirsk Region, Omsk State University, Tomsk State University, Tomsk Polytechnic University.

Banu Çiçek Başaran Uysal works at English Language Teaching Department at Eskisehir Osmangazi University, Turkey. She holds a bachelor's degree in Foreign Language Education from Middle East Technical University, with a minor in psychology. She received her Ph.D. in ELT, in 2020. During her doctoral studies, she conducted an extensive needs analysis study with in-service EFL teachers in Turkey and proposed an online in-service training model to address the identified needs based on her findings. Currently, she teaches various EFL teacher training courses at the undergraduate level including Teaching English to Young Learners, Practice Teaching, Second Language Acquisition, and Approaches to ELT courses. Her research interests encompass several areas within the field of language education. She is particularly interested in pre-service teacher education, in-service teacher training, technology integration, online education, and corpus studies.

Index

L

M

O

P

S

T

U

W

www.ingramcontent.com/pod-product-compliance
Lightning Source LLC
LaVergne TN
LVHW061536290926
849162LV00004B/816
* 9 7 8 1 6 6 8 4 6 2 2 2 5 *